SpringerBriefs in Earth Sciences

For further volumes:
http://www.springer.com/series/8897

Shunlin Liang · Xiaotong Zhang
Zhiqiang Xiao · Jie Cheng
Qiang Liu · Xiang Zhao

Global LAnd Surface Satellite (GLASS) Products

Algorithms, Validation and Analysis

 Springer

Shunlin Liang
Xiaotong Zhang
Jie Cheng
Qiang Liu
Xiang Zhao
State Key Laboratory of Remote
    Sensing Science
College of Global Change and Earth
    System Science
Beijing Normal University
Beijing
People's Republic of China

Zhiqiang Xiao
State Key Laboratory of Remote
    Sensing Science
School of Geography
Beijing Normal University
Beijing
People's Republic of China

Shunlin Liang
Department of Geographical Sciences
University of Maryland
College Park, MD
USA

ISSN 2191-5369   ISSN 2191-5377 (electronic)
ISBN 978-3-319-02587-2   ISBN 978-3-319-02588-9 (eBook)
DOI 10.1007/978-3-319-02588-9
Springer Cham Heidelberg New York Dordrecht London

Library of Congress Control Number: 2013950730

Printed on acid-free paper

Springer is part of Springer Science+Business Media (www.springer.com)

Acknowledgments

This book is based largely on the results of the "Generation and Application of Global Products of Essential Land Variables" project, which was funded and managed by the National Remote Sensing Center of China, Ministry of Science and Technology of China (Grant No. 2009AA122100), along with the participation of researchers from more than 20 universities and research institutes in China. Prof. Guanhua Xu, Academician of the Chinese Academy of Sciences and former Minister of Science and Technology of China, provided consistent support and advice. Without his encouragement and guidance, this project would not have been possible. In addition, we would like to express our deepest thanks for the support and guidance from many experts around the world.

We thank the SURFRAD, AERONET, and CarbonEuropeIP programs as well as the principal investigators of the selected sites where the ground measurement data were provided for our validation. We would also like to thank NASA/EOS for providing the MODIS and AVHRR data, NOAA for the GOES data, EUMESAT for the MSG SEVIRI data, and JMA for the MTSAT data. This work was also supported by the Natural Science Foundation of China (No. 40901167 and 41371323, 41101310).

We especially thank our colleagues, friends, and particularly our families who have provided enormous support throughout this project.

Contents

Chapter 1
Introduction

The inhabitants of planet Earth are currently facing unprecedented environmental challenges, such as shortages of clean and accessible freshwater, degradation of terrestrial and aquatic ecosystems, and increases in extreme climatic events. Remote sensing technology using Earth-orbiting satellites is one of the most effective tools for providing data to help researchers address these global environmental problems. Since the 1970s, hundreds of Earth observation satellites have been launched, and observing Earth from space has fundamentally transformed the way we view and study our home planet. Remote sensing observations have resulted in new discoveries, transformed the discipline of Earth sciences, opened new avenues of research, and improved the predictability of Earth system processes such as hurricanes, tornados, floods, and droughts.

Satellite remote sensing technologies have grown increasingly more sophisticated in recent decades, evolving from simple photographs to highly detailed quantitative measurements of the Earth's biogeophysical and biogeochemical properties. Transforming raw satellite data into Earth properties requires a series of data processing, inversion, and analytical operations, which cannot practically be carried out by individual application users and should rather be executed by specialists using centralized facilities. As a result, data centers are distributing more and more high-level satellite products rather than simply raw satellite imagery.

To achieve a better understanding of global environmental changes, a long-term, high-resolution, and high-quality global dataset of land surface products is needed. We have undertaken a project known as the "Generation and Applications of Global Products of Essential Land Variables," which is supported by China's National High Technology Research and Development Program, the "863 program." Vast amounts of remotely sensed data have been collected from more than a dozen satellites and ground measurements. From these data, a suite of new inversion algorithms and five Global LAnd Surface Satellite (GLASS) products have been developed. Table 1.1 summarizes the basic characteristics of these five GLASS products.

S. Liang et al., *Global LAnd Surface Satellite (GLASS) Products*,
SpringerBriefs in Earth Sciences, DOI: 10.1007/978-3-319-02588-9_1,
© The Author(s) 2014

Table 1.1 Characteristics of the five GLASS products

Product	Spatial resolution	Temporal resolution	Temporal range
Shortwave albedo	1–5 km, 0.05°	8-day	1981–2012
Incident solar radiation	5 km, 0.05°	3-h	2008–2010
Incident PAR	5 km, 0.05°	3-h	2008–2010
Longwave emissivity	1–5 km, 0.05°	8-day	1981–2012
LAI	1–5 km, 0.05°	8-day	1981–2012

These GLASS products are unique. LAI, shortwave albedo, and longwave emissivity are spatially continuous and cover the longest period of time among all current similar satellite products. The two global radiation products at 5 km spatial resolution and 3 h temporal resolution from 2008 to 2010 are the highest resolution global radiation products from satellite observations in the world. These products seamlessly record the changes occurring on the global land surface over the past 30 years, thus providing climate-change researchers worldwide with valuable data and information on the global land-surface energy budget and on water and carbon cycles.

The scope of the overall project and the GLASS product production system has been described elsewhere (Liang et al. 2013; Zhao et al. 2013). To facilitate and promote the use of these five products, details of the algorithms and their characteristics are provided in the subsequent chapters.

References

Liang S, Zhao X, Yuan W, Liu S, Cheng X, Xiao Z, Zhang X, Liu Q, Cheng J, Tang H, Qu YH, Bo Y, Qu Y, Ren H, Yu K, Townshend J (2013) A long-term global land surface satellite (GLASS) data-set for environmental studies. Int J Digital Earth. doi: 10.1080/17538947.17532013.17805262

Zhao X, Liang S, Liu S, Yuan W, Xiao Z, Liu Q, Cheng J, Zhang X, Tang H, Zhang X, Liu Q, Zhou G, Xu S, Yu K (2013) The global land surface satellite (glass) remote sensing data processing system and products. Remote Sens 5:2436–2450

Chapter 2
Leaf Area Index

Abstract This chapter briefly introduces the inversion algorithms used, discusses the product characteristics and validation results, and presents preliminary analyses and applications. The inversion algorithm was developed to estimate LAI from time-series remote sensing data using general regression neural networks (GRNNs). Unlike existing neural network methods that use remote sensing data acquired only at a specific time to retrieve LAI, the GRNNs used in this study are trained using the fused time-series LAI values from the MODIS and CYCLOPES LAI products and the reprocessed MODIS/AVHRR reflectance. The reprocessed time-series MODIS/AVHRR reflectance values from an entire year were input to the GRNNs to estimate the 1 year LAI profiles. This algorithm has been used to produce the GLASS LAI, one of the longest duration (1981–2012) LAI products in the world. Extensive validations for all biome types have been carried out, and it has been demonstrated that the GLASS LAI presents temporally continuous LAI profiles with much improved quality and accuracy compared with those of the current MODIS and CYCLOPES LAI products.

Keywords Leaf area index · General regression neural network · MODIS · AVHRR · Retrieval · Time series · GLASS

2.1 Background

Leaf area index (LAI) is defined as one half the total green leaf area per unit of horizontal ground surface area and is called *true LAI*. The true LAI multiplied by the clumping index is called *effective LAI*. LAI measures the amount of leaf material in an ecosystem, which imposes important controls on processes such as photosynthesis, respiration, and rain interception that link vegetation to climate. Hence, LAI appears as a key variable in many models that describe vegetation-atmosphere interactions, particularly with respect to the carbon and water cycles.

S. Liang et al., *Global LAnd Surface Satellite (GLASS) Products*,
SpringerBriefs in Earth Sciences, DOI: 10.1007/978-3-319-02588-9_2,
© The Author(s) 2014

Models typically use satellite-based estimates of LAI in one of three ways: model forcing, validation of model output, and model assimilation. Buermann et al. (2001), using the National Center for Atmospheric Research (NCAR) Community Climate Model (CCM3) forced with an AVHRR-derived LAI, reported that the use of satellite derived fields revealed substantial warming and decreased precipitation over large parts of the Northern Hemisphere land areas during the boreal summer. This warming and drying reduced the discrepancies between simulated and observed near-surface temperature and precipitation fields.

The LAI user community includes the following categories (Myneni 2007): (a) scientific users: modelers of climate, primary production, ecology, hydrology, and crop production; (b) public users: meteorological organizations, deforestation and desertification monitoring organizations, rapid-response systems, pest-risk evaluation companies, governments involved in implementation of international treaties such as the Kyoto Protocol; and (c) private users: international agriculture and forestry companies, insurance companies, and traders.

The World Meteorology Organization (WMO) requirements shown in Tables 2.1 and 2.2 summarize the current global LAI products. Comparisons of these two tables indicate that current products do not meet user requirements.

Based on feedback from the user community and accumulated research experience, Myneni et al. (2007) proposed the following specifications for a global LAI product in a NASA Earth Science Data Record (ESDR) White paper:

Table 2.1 WMO observation requirements for LAI by space program. (http://www.wmo-sat.info/db/variables/view/98, updated June 3, 2012)

Application area	Uncertainty goal (%)	Uncertainty threshold (%)	Spatial resolution goal (km)	Spatial resolution threshold (km)	Temporal resolution goal	Temporal resolution threshold (days)
Global NWP	5	20	2	50	24 h	10
High-resolution NWP	5	20	1	40	12 h	2
Hydrology	5	20	0.01	10	7d	24
Agricultural meteorology	5	10	0.01	10	5d	7
Climate-TOPC	5	10	0.25	10	24 h	30

Table 2.2 Characteristics of current global LAI products

LAI products	LAI type	Spatial resolution	Temporal resolution (days)	Temporal range
MODIS	True	1 km	8	2000–present
CYCLOPES	Effective	1/112°	10	1999–2007
GLOBECARBON	True	1/11.2°	10	1998–2007
Geoland2	True	1 km, 0.05°	10	1999–2012
GLASS	True	1–5 km, 0.05°	8	1981–2012

- Accuracy of 0.5 LAI units to be achieved for corresponding global averages over individual biomes;
- Spatial resolution, depending on application from 250 m (local ecological studies) to 0.25 degree (global climate studies);
- Temporal frequency from 4 days to monthly;
- Length of recording starting from the beginning of AVHRR measurements (July 1981) and continuing into the future.

2.2   Algorithms

There are two methods for retrieving LAI from satellite data (Liang et al. 2012; Liang 2007): empirical methods and physical methods. The empirical methods are based on statistical relationships between LAI and spectral vegetation indices, which are calibrated for distinct vegetation types using field measurements of LAI and reflectance data recorded by a remote sensor or simulations with canopy radiation models. The physical methods are based on the inversion of canopy radiative transfer models through iterative minimization of a cost function, the look-up table (LUT) method, or various machine learning methods. Inversion techniques based on iterative minimization of a cost function require hundreds of runs of the canopy radiative transfer model for each pixel and are therefore computationally too demanding. For operational applications, LUT and artificial neural networks (ANN) methods are two popular inversion techniques that are based on a precomputed reflectance database. For example, MODIS and MISR LAI products are based on the LUT method (Myneni et al. 2002) and the CYC-LOPES LAI product is based on the ANN method (Baret et al. 2007).

The GLASS LAI retrieval algorithm (Xiao et al. 2013) uses general regression neural networks (GRNNs) to retrieve LAI values from time-series MODIS/AVHRR reflectance data. The GRNNs are trained using fused LAI samples from MODIS and CYCLOPES LAI products and the reprocessed MODIS/AVHRR reflectance of the BELMANIP sites during the period from 2001 to 2004.

Two unique features of the GLASS LAI algorithm should be emphasized here. The first is that the LAI annual profile is estimated using annual observations. Unlike existing neural network methods that use remote sensing data acquired only at a specific time to retrieve LAI, the GRNNs used in the GLASS LAI production use the surface reflectance of an entire year as their input. The output is a 1 year LAI profile for each pixel. The second feature is that the GRNNs are training by integrating both MODIS and CYCLOPE LAI products. Training with representative samples is of critical importance in any ANN algorithms. Instead of using simulation data, the GRNNs are trained using fused time-series LAI values from MODIS and CYCLOPES LAI products and the reprocessed MODIS/AVHRR reflectance. A flow chart outlining this method is shown in Fig. 2.1.

A database was generated from the MODIS and CYCLOPES LAI products and the MODIS/AVHRR reflectance products at the BELMANIP sites from 2001 to

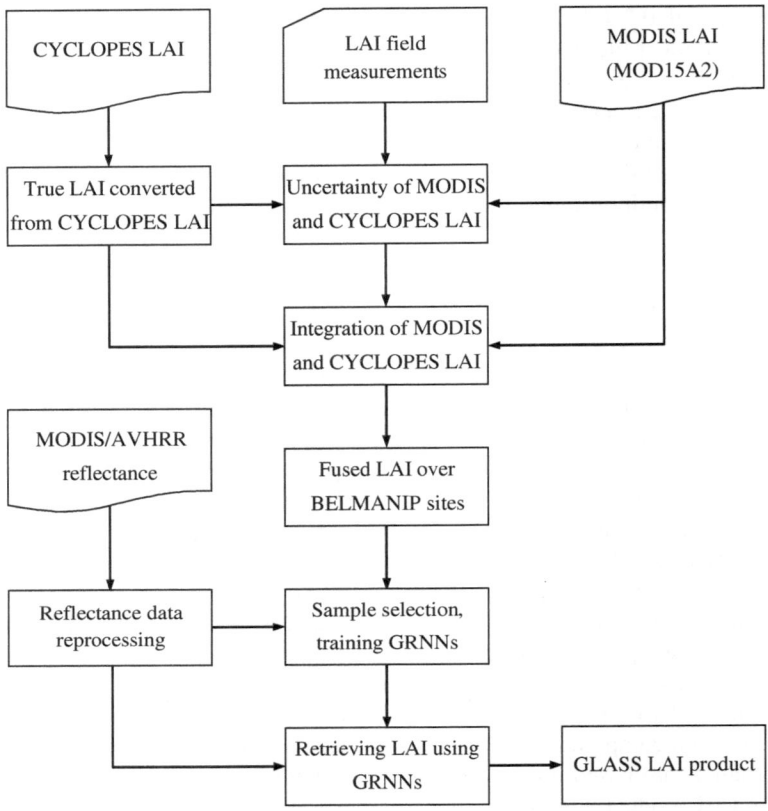

Fig. 2.1 Flow chart of LAI retrieval using general regression neural networks. (Xiao et al. 2013), Copyright © 2003 reproduced with permission of IEEE

2004. The effective CYCLOPES LAI was first converted to true LAI and then combined with the MODIS LAI according to their uncertainties determined from the ground-measured true LAI values. The MODIS and AVHRR reflectance values were reprocessed to remove any remaining effects from cloud contamination and other factors. The GRNNs were then trained using the fused LAI and the reprocessed MODIS/AVHRR reflectance for each biome type. To retrieve LAI values from time-series remote sensing data, the reprocessed MODIS/AVHRR reflectance data from an entire year were input to estimate one-year LAI profiles using the GRNNs. A detailed description of the new algorithm is given in the following subsections.

2.2.1 General Regression Neural Networks

The GRNN, developed by Specht (1991), is a generalization of radial basis function networks (RBFNs) and probabilistic neural networks (PNNs). The advantage of this type of neural network is that it can approximate the map inherent in any sample data set. In addition, GRNNs possess a special property that these networks do not require iterative training. The functional estimate is computed directly from the training data. At present, GRNNs have been applied in a variety of fields, such as system identification, adaptive control, pattern recognition, and time series prediction (Leung et al. 2000).

Figure 2.2 shows a GRNN with a multi-input–multi-output architecture. It includes four layers: the input layer, the pattern layer, the summation layer, and the output layer. The input layer provides all the measurement variables to all the neurons in the pattern layer; each neuron represents a training pattern, and the output of each neuron is a measure of the distance of the input from the stored patterns. The summation layer consists of two types of summation neurons: one type computes the summation of the weighted outputs of the pattern layer, where the weight for the ith neuron in the pattern layer is the target output value corresponding to the ith input pattern, and the other type calculates the unweighted outputs of the pattern neurons. Finally, the output layer performs a normalization step to compute the GRNN-predicted value of the output variable.

If the kernel function of the GRNN is Gaussian, the fundamental formulation of the GRNN can be deduced as follows:

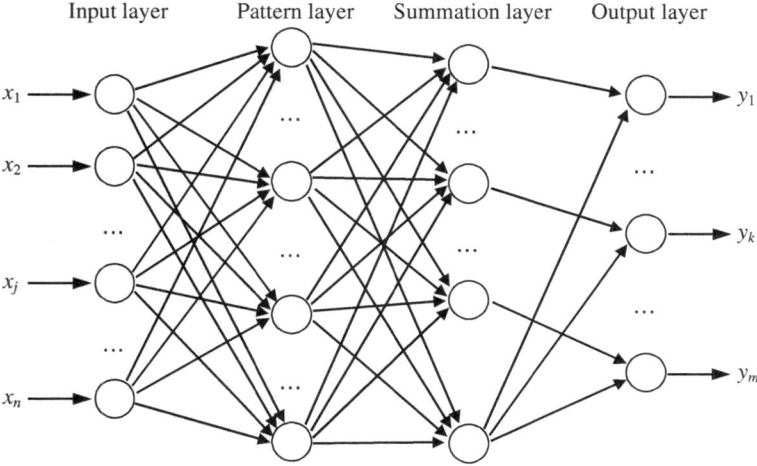

Fig. 2.2 GRNN with a multi-input–multi-output architecture. (Xiao et al. 2013), Copyright © 2003 reproduced with permission of IEEE

$$Y'(X) = \frac{\sum\limits_{i=1}^{n} Y^i \exp\left(-\frac{D_i^2}{2\sigma^2}\right)}{\sum\limits_{i=1}^{n} \exp\left(-\frac{D_i^2}{2\sigma^2}\right)}, \tag{2.1}$$

where $D_i^2 = (X - X^i)^T (X - X^i)$ represents the squared Euclidean distance between the input vector $\mathbf{X}$ and the ith training input vector $\mathbf{X}^i$, Y^i is the output vector corresponding to the vector $\mathbf{X}^i$, $Y'(X)$ is the estimate corresponding to the vector $\mathbf{X}$, n is the number of samples, and σ is a smoothing parameter that controls the size of the receptive region. As σ becomes larger, the GRNN output approaches the mean of the training set outputs. As σ becomes smaller, the GRNN output approaches the output pattern of the training set. Equation (2.1) shows that the estimate $Y'(X)$, given an input vector $\mathbf{X}$, is the weighted average of all the sample observations $\mathbf{Y}^i$, where the weight for each observation is proportional to the Euclidean distance between the vector $\mathbf{X}$ and the training input vector $\mathbf{X}^i$.

The input vector $\mathbf{X}$ of the GRNNs used to retrieve LAI includes the reprocessed MODIS/AVHRR time-series reflectance values in the red (R) and near-infrared (NIR) bands (for a one-year period); that is, $X = (R_1, R_2, \ldots, R_{46}, \mathrm{NIR}_1, \mathrm{NIR}_2, \ldots, \mathrm{NIR}_{46})^T$ and contains 92 components. The output vector $Y' = (\mathrm{LAI}_1, \mathrm{LAI}_2, \cdots, \mathrm{LAI}_{46})^T$ is the corresponding LAI time series for the year and contains 46 components.

2.2.2 Generating the Training Database

To generate LAI products at the regional to global scale using GRNNs, the training database should be globally representative of surface types and conditions. The BELMANIP network, which includes 402 sites, aims to provide a good sampling of biome types and conditions throughout the world (Baret et al. 2006). For each BELMANIP site, a 3×3 subset of the MODIS and CYCLOPES LAI products and the MODIS reflectance products for 2001–2004 was extracted.

The MODIS LAI product represents a true LAI, while the CYCLOPES LAI product represents an effective LAI. Therefore, there is a great need for a conversion relationship between effective and true LAI for use when integrating multiple LAI products.

Chen et al. (1996) demonstrated that the effective LAI (LAI_e) can be determined as the product of the true LAI (LAI_t) and the clumping index (Ω) as follows:

$$\mathrm{LAI}_e = \Omega \times \mathrm{LAI}_t \tag{2.2}$$

Based on the linear relationship between the clumping index and the normalized difference between hotspot and darkspot (NDHD) indices, Chen et al. (2005) derived the first global clumping index map using multiangular POLDER 1 satellite data from ADEOS-1. Pisek et al. (2010) expanded the global mapping of the clumping index by integrating new, complete, year-round observations from POLDER 3. The across-biome difference in the topographical effect was removed in the new global clumping index map. The spatial resolution of the global clumping index map is 6 km.

Global average of the clumping index values for different biome types were derived according to the MODIS land cover product. For each land cover class, the pixels in the global clumping index map were selected to calculate the global average if at least 85 % of the pixels were covered with a single MODIS land cover type. In this case, the CYCLOPES LAI for different vegetation types can be converted to the true LAI using Eq. (2.2).

After the CYCLOPES LAI was converted to true LAI, the true CYCLOPES and MODIS LAI values were combined according to their uncertainties as determined from the ground-measured true LAI.

The fused LAI was generated as follows:

$$\text{LAI}_{\text{modcyc}} = w_{\text{mod}}\text{LAI}_{\text{mod}} + w_{\text{cyc}}\text{LAI}_{\text{cyc}}^{*}, \qquad (2.3)$$

where $\text{LAI}_{\text{modcyc}}$ is a combined estimate of LAI, LAI_{mod} is the smoothed and gap-filled MODIS LAI resulting from the multistep Savitzky-Golay filtering procedure (Xiao et al. 2011), $\text{LAI}_{\text{cyc}}^{*}$ is the true LAI converted from the CYCLOPES LAI using the method described above, and w_{mod} and w_{cyc} are the normalized weights for the MODIS LAI and the true CYCLOPES LAI, respectively, the sum of these weights being equal to one.

By this method, the fused LAI can be obtained using Eq. (2.3) if the weights are known. Figure 2.3 shows the fused LAI time series for four different biome types. For convenient comparison, the MODIS and CYCLOPES LAIs and their corresponding weights are also shown in Fig. 2.3. Because the true CYCLOPES LAI is more consistent with the ground-measured LAI than is the MODIS LAI, the CYCLOPES weights are larger than the MODIS weights during the growing season.

The MODIS and AVHRR reflectance values were then processed further. The quality of MODIS reflectance is affected by many factors such as clouds, aerosols, water vapor, and ozone. Although many of these effects can be removed through atmospheric corrections, the remaining effects can sometimes be very large and require further processing (Chen et al. 2006). Much effort has been devoted to the development of reprocessing methods to remove residual atmospheric contamination from the reflectance. In this study, the method developed by Tang et al. (2013) was used for MODIS reflectance reprocessing. Data contaminated by undetected and fallout clouds were identified in the MODIS reflectance data using MODIS snow and cloud mask data, the spectral characteristics of temporal and spatial continuity and correlation, and other auxiliary information. Then, the cloud-contaminated data were removed using temporal-spatial filtering algorithms, and the missing data were filled in using an optimum interpolation algorithm to obtain smooth continuous surface reflectance values. Detailed information on this reprocessing method for MODIS reflectance is given by Tang et al. (2013).

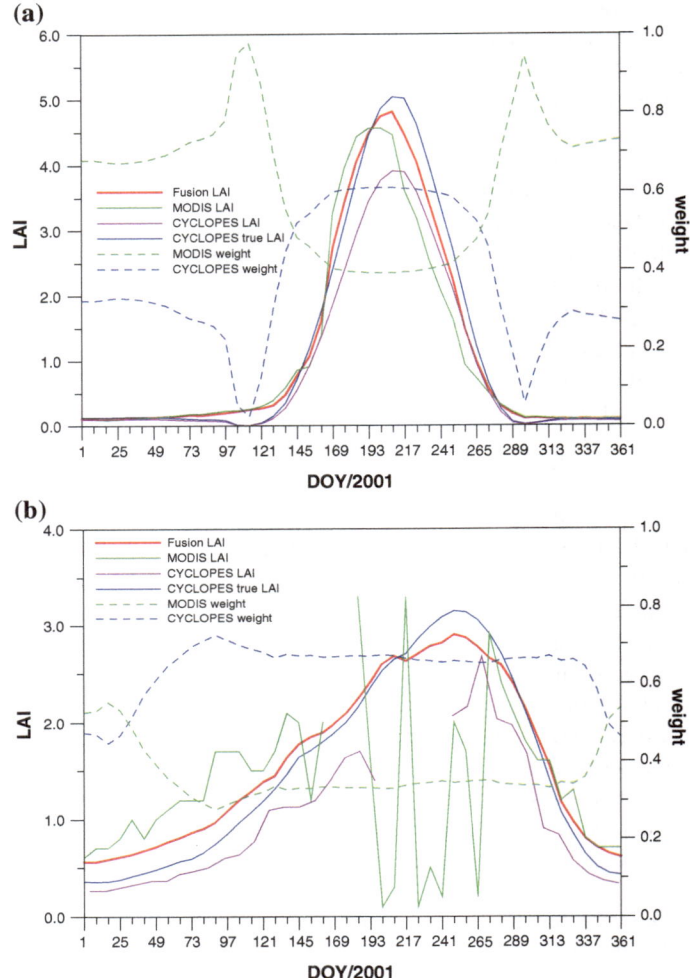

Fig. 2.3 Fused LAI time series from MODIS and CYCLOPES LAI values for different vegetation types: **a** crops, **b** grasses, **c** broadleaf forests, and **d** coniferous forests. (Xiao et al. 2013), Copyright © 2003 reproduced with permission of IEEE

Figure 2.4 shows the reprocessed reflectance in the red (R) and near-infrared (NIR) bands at the Bondville site. It can be observed that the reprocessed reflectance is relatively smooth compared to the original MODIS reflectance. The fused time-series LAI and the reprocessed MODIS reflectance values were then used to train the GRNNs.

Similarly, to the MODIS surface reflectance data, the AVHRR reflectance data were also reprocessed to remove cloud-contaminated data and fill in the missing data using the method developed by Tang et al. (2013). In addition, the maximum

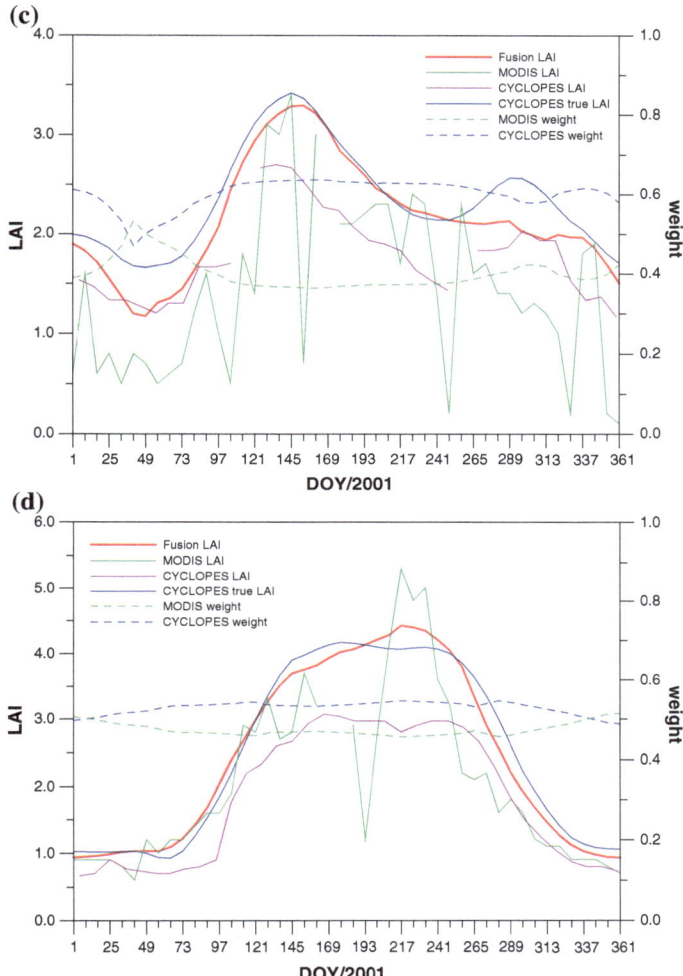

Fig. 2.3 continued

value composite approach (MVC) was used to combine the daily surface reflectance data into composites of eight-day intervals to maintain a consistent time resolution with MODIS surface reflectance data (Liang et al. 2012). The MVC method selects the AVHRR reflectance data with the highest normalized difference vegetation index (NDVI) over each eight-day time interval. To generate training database of the GRNNs for retrieval of LAI from time series AVHRR reflectance data, the fused LAI values at the BELMANIP sites calculated from MODIS, and CYCLOPES LAI products for 2003–2004 were transformed to the geographic projection and aggregated to 0.05° resolution using a spatial average sampling

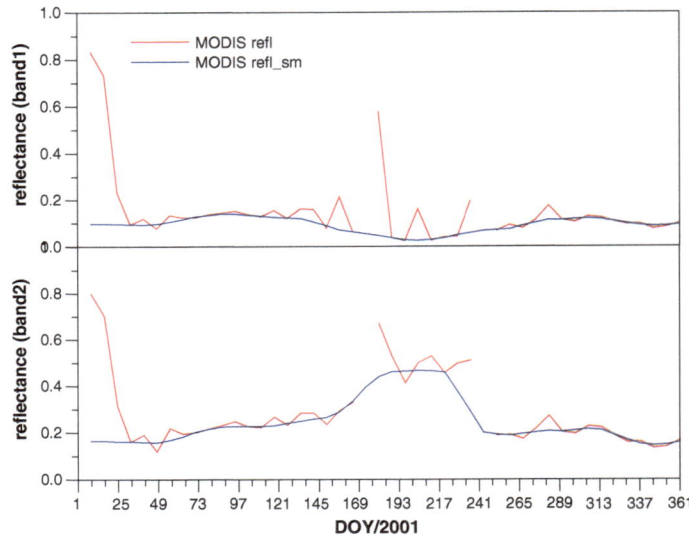

Fig. 2.4 Reprocessed reflectance (MODIS refl_sm) and original MODIS reflectance (MODIS refl) in the R and NIR bands. (Xiao et al. 2013), Copyright © 2003 reproduced with permission of IEEE

method. Then, the fused time-series LAI values were used to train GRNNs with the corresponding reprocessed AVHRR reflectance values.

To achieve better performance and convergence, the values of the training inputs and outputs were normalized according to Eq. (2.4) before training the GRNNs:

$$X_{norm} = 2.0 \times (X - X_{min})/(X_{max} - X_{min}) - 1, \qquad (2.4)$$

where X_{max} and X_{min} are the maximum and minimum values, for variable X, and X_{norm} is the normalized value corresponding to the training inputs or outputs X.

2.2.3 Training of the GRNNs

Unlike back-propagation neural networks, which are iteratively trained to determine the weights, the architecture and weights of GRNNs are determined when the input to the GRNN is given. The smoothing parameter σ is the only free parameter in the GRNN formulation. Therefore, training of a GRNN is essentially optimization of the smoothing parameter. The value of the smoothing parameter significantly affects the accuracy of the GRNN predictions. Therefore, the magnitude of σ must be chosen carefully.

Specht (1991) suggested the use of the holdout method to find a suitable σ. For a particular value of σ, the holdout method consists of removing one sample at a time from the training dataset and constructing a GRNN based on all the other training samples. The GRNN is then used to estimate Y for the removed sample. By repeating this process for each sample and storing each estimate, the mean squared error between the actual sample values $\mathbf{Y}^i$ and the estimates can be evaluated. The value of σ giving the smallest error should be used in the final GRNN. In fact, for the GRNNs in this study, the learning period was completed when the minimum of the following cost function of the smoothing parameter was reached:

$$f(\sigma) = \frac{1}{n}\sum_{i=1}^{n}\left[\hat{Y}_i(X_i) - Y_i\right]^2, \qquad (2.5)$$

where $\hat{Y}_i(X_i)$ is the estimate corresponding to X_i using the GRNN trained over all the training samples except the ith sample.

A variety of optimization methods are currently available to find the optimal smoothing parameter in Eq. (2.5). The most commonly used methods are the hill-climbing method and the conjugate gradient method. However, these are prone to become caught in local minima, and they can produce false minima. Hansen and Meservy (1996) used a genetic algorithm to optimize the smoothing factor to find the optimal GRNN regression surface for global optimization. In this study, the shuffled complex evolution method developed at the University of Arizona (SCE-UA) was used to obtain the optimal smoothing parameter for the GRNN; this algorithm does not require the derivatives of the function, and is not susceptible to being trapped by small pits and bumps on the function surface (Duan et al. 1992). It has been extensively used in our recent studies (Xiao et al. 2009, 2012).

Using the MODIS land cover product, a dedicated GRNN was constructed for each biome type. Then, the trained GRNNs were used to retrieve LAI for the corresponding biome types from the reprocessed MODIS and AVHRR reflectance values.

2.3 Product Characteristics and Quality Control

2.3.1 GLASS LAI Product Characteristics

The GLASS LAI product, archived in the HDF-EOS format, has a temporal resolution of 8 days and is available from 1981 to 2012. From 2000 to 2012, the LAI product is derived from MODIS reflectance. It is provided in an Integerized Sinusoidal (ISIN) projection at a resolution of 1 km for each of the 289 land tiles (each representing 1200×1200 pixels) on the Earth. From 1981 to 1999, the LAI

(a)   **(b)**

(c)   **(d)**

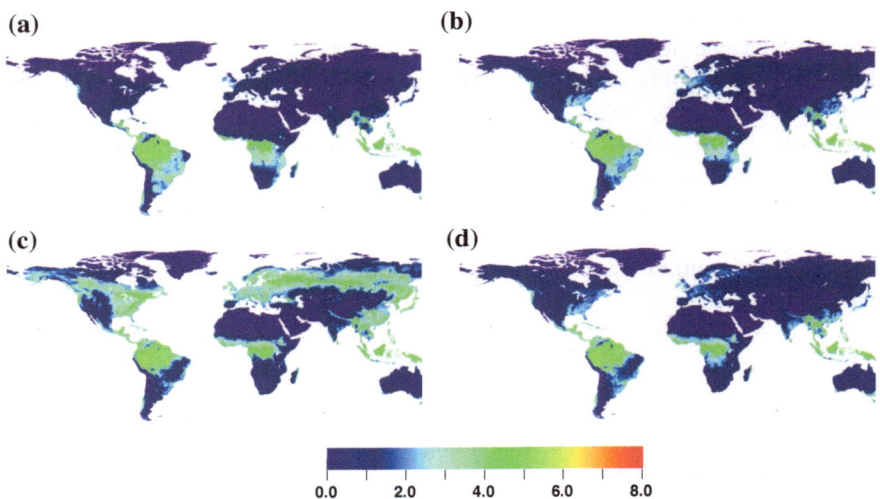

0.0   2.0   4.0   6.0   8.0

Fig. 2.5 GLASS LAI maps for January (**a**), April (**b**), July (**c**), and October (**d**) 2005. Geographic projection, 0.05° resolution for display purposes

product was generated from AVHRR reflectance. It is provided in a geographic projection at a resolution of 0.05°.

Four examples of the GLASS LAI global product for January, April, July, and October 2005 (Fig. 2.5) show a coherent global spatial distribution of the GLASS LAI product, presenting the highest values over equatorial forest, and zero values in desert areas.

The GLASS LAI product is spatially complete, and no gaps are present even if the MODIS/AVHRR surface reflectance is contaminated due to cloud or missing due to sensor failure.

Using the basic data product described above, 0.05° and 1° spatial resolution products were produced from 1981 to 2012 in the geographic projection of the Climate Modeling Grid (CMG) to meet the needs of global change and climate studies. The CMG products are also available for distribution.

2.3.2 Quality Control

Every scene of the GLASS LAI products has been subjected to a strict quality control (QC) procedure. Quality control was carried out both automatically and with human involvement. Automatic QC refers mainly to the generation of a QC flag that can be used to estimate the uncertainty in the GLASS LAI products. The quality check with human involvement is a computer-aided visual inspection of the spatial and temporal variations in the LAI product before its final release.

Table 2.3 Quality control flags of the GLASS LAI product

Bit	Bit combination	Meaning
0–1	00	LAI value with high quality
	01	LAI value with good quality
	10	Fill value
2–3	00	LAI value is retrieved using GRNNs
	01	Fill value
4	0	Non-water
	1	Water
5	0	Non-snow
	1	Snow
6	0	Non-cloud
	1	Cloud
7	0	The cloud has not been processed, and original value is retained
	1	The cloud has been processed

The term "LAI" is referred to an LAI value stored in an eight-bit signed integer data type. Valid product values range from 0 to 100, and the scale factor that converts the digital number to an LAI value is 0.1. The pixels representing water are filled with a digital number of 255. "QC" is the quality control flag, which gives a pixel-wise description of the data processing parameters as well as an indication of credibility of the result. The flag is an eight-bit unsigned integer data type provided for each pixel. The bitwise interpretation of the QC flags for the GLASS LAI product is given in Table 2.3.

2.4 Product Validation

To evaluate the accuracy of the GLASS LAI product, its results were first compared with those from other global LAI products, in addition, a direct validation was also performed (Xiang et al. 2013).

2.4.1 Cross-Comparison of GLASS LAI with Other Global LAI Products

A cross-comparison and validation of the GLASS, MODIS, and CYCLOPES LAI products was performed. The MODIS LAI product has been produced since 2000 at a 1-km spatial resolution and an 8-day time step. The MODIS LAI retrieval algorithm includes a main algorithm and a backup algorithm. The main algorithm is based on LUTs simulated from a three-dimensional radiative transfer model for eight main biome classes. When the main algorithm fails, the backup algorithm is used to estimate LAI from biome-specific LAI-NDVI relationships (Knyazikhin et al. 1998). The CYCLOPES LAI product, with a spatial resolution of $1/112°$ and a 10-day temporal sampling interval, was generated from SPOT/VEGETATION sensor data for 1999–2003 (Weiss et al. 2007). The algorithm used to estimate LAI was based on training neural networks with PROSPECT + SAIL radiative transfer model simulations (Weiss et al. 2007). The CYCLOPES LAI product was projected in *plate carrée*. Because of the different projection systems used for the GLASS, MODIS, and CYCLOPES LAI products, the CYCLOPES LAI values were reprojected into the MODIS projection system using the General Cartographic Transformation Package (GCTP) map projections library (U.S. Geological Survey 1993), and resampled to exhibit a 1 km spatial resolution using a bilinear interpolation technique.

2.4.1.1 Spatial Consistency

Results from global comparisons of the GLASS, MODIS, and CYCLOPES LAI products on DOY 25 and 217 in 2003 are presented in this section. Figure 2.6 shows the spatial patterns of these LAI products at a global scale. It can be observed that the GLASS LAI was consistent with the MODIS and CYCLOPES LAI products and in rough agreement as to magnitude. The highest LAI values and the relatively large differences between products were observed at latitudes 15°S–15°N, close to the Equator, and near 50°N in Russia and Canada. The CYCLOPES LAI maintained the lowest values and was never greater than 4 on DOY 217 in most regions of South America close to the Equator. In the same area, the GLASS and MODIS LAI values were between 4 and 6.

Better spatial completeness was achieved by GLASS and MODIS than by CYCLOPES. High-latitude regions in the Northern Hemisphere were blank due to the effects of snow on DOY 25. On DOY 217, as shown in the right-hand column in Fig. 2.6, GLASS and MODIS had almost no fill values anywhere in the world.

All the products followed the seasonality effect, which shows opposite properties in the two hemispheres. In the Northern Hemisphere (NH), the LAI values in the right-hand column (summer) were much higher than those in the left-hand column (winter). In fact, in the left-hand column, the LAI values for the Northern Hemisphere were between 0 and 1, and the spatial variation was very low globally. There were many missing data points, represented by blanks, at high latitudes due to snow in winter. In the right-hand column, the Northern Hemisphere showed large spatial variations, with high LAI values in Canada and Russia. Similarly, in the Southern Hemisphere (SH), the LAI values in summer were larger than in winter. The spatial inconsistency in the SH was greater than in the NH.

To analyze more deeply the differences between GLASS and other LAI products, maps of their frequency distributions of differences were produced (see Fig. 2.7). The GLASS LAI values were in good agreement with the MODIS LAI values with the best possible quality (QC < 32). When the QC of the MODIS LAI values was greater than 32, large discrepancies were observed between the GLASS LAI values and the MODIS LAI values. The properties of the differences between

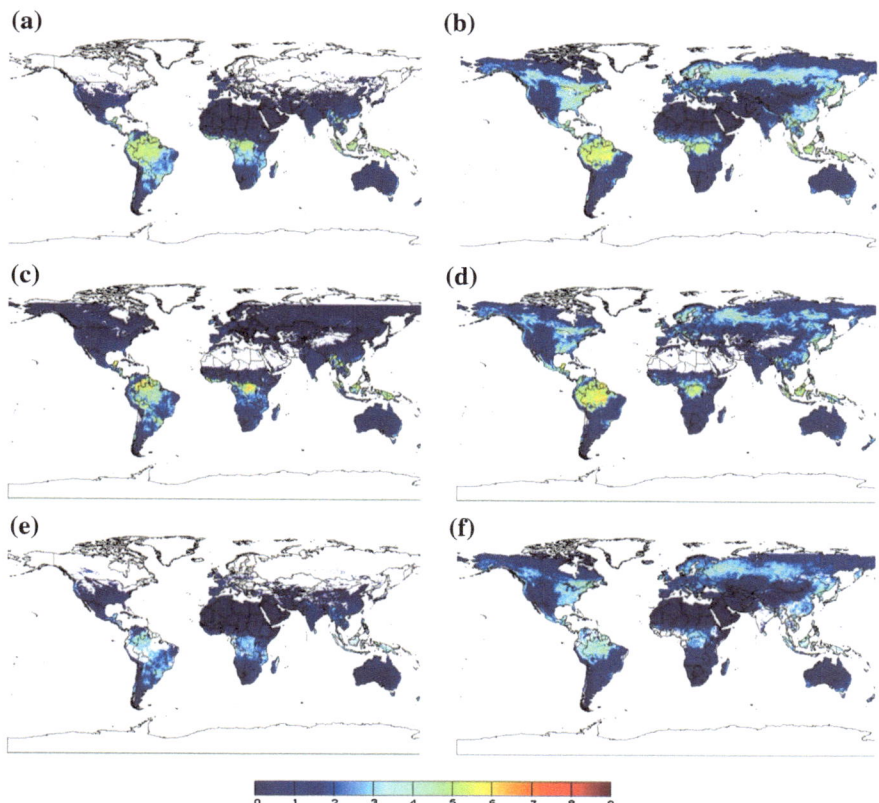

Fig. 2.6 Spatial patterns of the GLASS, MODIS, CYCLOPES, and CCRS LAI products on DOY 25 and 217 of 2003. Fill values are set to blanks. **a** GLASS LAI, DOY 25, 2003, **b** GLASS LAI, DOY 217, 2003, **c** MODIS LAI, DOY 25, 2003, **d** MODIS LAI, DOY 217, 2003, **e** CYCLOPES LAI, DOY 25, 2003, **f** CYCLOPES LAI, DOY 217, 2003

GLASS and MODIS described above suggest that the GLASS LAI has improved upon the unrealistically high or low values of the MODIS LAI when the QC is poor, especially for forest.

Figure 2.8 shows a histogram of the discrepancy map of the GLASS and CYCLOPES LAI values on DOY 217 in 2003. Because of the mismatch between the projections used in the GLASS and CYCLOPES products, the projection of the CYCLOPES LAI was converted into sinusoidal form using the GCTP library. Then the difference values were computed by subtracting the reprojected CYCLOPES LAI from the GLASS LAI. Better consistency was achieved between the GLASS and CYCLOPES products than between GLASS and MODIS products. The GLASS LAI was larger than the CYCLOPES LAI by approximately −0.5 to 0.5 over most of the study regions. The histogram had only one peak near zero, and the percentage at the zero value was 27 %. More than 70 % of the difference

Fig. 2.7 Histograms of the
discrepancies between
GLASS and MODIS LAI

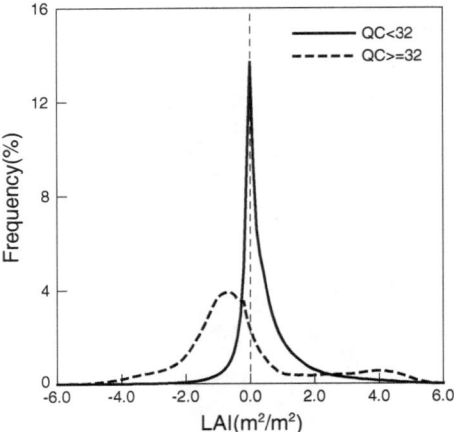

Fig. 2.8 Histogram of
differences between GLASS
and CYCLOPES LAI values
on DOY 217 in 2003

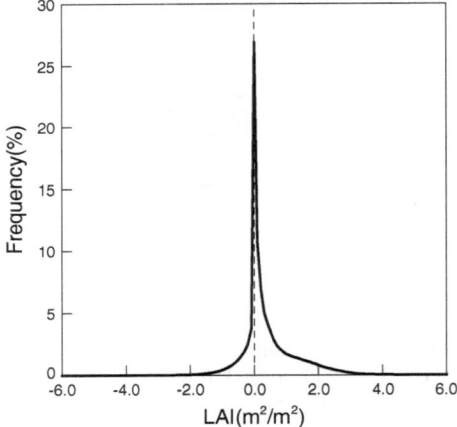

values were between −0.5 and 0.5. The number of positive discrepancies was far greater than the number of the negative discrepancies because of underestimation by the CYCLOPES algorithm. Moreover, larger discrepancy values, between 0.5 and 3, were observed near the Equator and in the higher latitudes of the Northern Hemisphere. These large discrepancies occurred because of underestimation by the CYCLOPES LAI in forest areas, due to early saturation of the CYCLOPES LAI (values are never larger than 4) and the lack of a clumping representation (Weiss et al. 2007).

Table 2.4 Characteristics of the validation sites (total 19)

Site	Country	Latitude (°)	Longitude (°)	DOY	Year	Mean LAI
Alpilles	France	43.8104	4.7146	204	2002	1.69
Camerons	Australia	−32.5983	116.2542	63	2004	2.13
Demmin	Germany	53.8921	13.2072	164	2004	4.15
Donga	Benin	9.7701	1.7784	172	2005	1.85
Fundulea	Romania	44.4058	26.5849	144	2002	1.53
Gilching	Germany	48.0819	11.3205	199	2002	5.39
Gnangara	Australia	−31.5339	115.8824	61	2004	1.01
Larose	Canada	45.3805	−75.2170	219	2003	5.87
Larzac	France	43.9375	3.1230	183	2002	0.81
Nezer	France	44.5680	−1.0382	107	2002	2.38
Plan-de-Dieu	France	44.1987	4.9481	189	2004	1.13
Puechabon	France	43.7246	3.6519	164	2001	2.85
Sonian	Belgium	50.7682	4.4111	174	2004	5.66
Sud-Ouest	France	43.5063	1.2375	189	2002	1.96
Wankama	Niger	13.6450	2.6353	174	2005	0.14
Zhangbei	China	41.2787	114.6878	221	2002	1.26
Counami	French Guyana	5.3435	−53.2368	269	2001	4.93
				286	2002	4.37
Laprida	Argentina	−36.9904	−60.5527	311	2001	5.82
				292	2002	2.81
Turco	Bolivia	−18.2350	−68.1836	208	2001	0.31
				240	2002	0.04

2.4.1.2   Temporal Consistency

The LAI temporal profiles of the central pixel at various sites in Table 2.4 are depicted in Figs. 2.9–2.13. To enable better evaluation of the quality of the GLASS LAI values, the temporal profiles of CYCLOPES LAI values, MODIS LAI values, and ground measurements are also shown for each selected site.

Figure 2.9 shows the temporal LAI trajectories over the Alpilles site for 2002. The biome type is broadleaf crops according to the MODIS land cover map. The GLASS, and CYCLOPES LAI values are in very good agreement for the entire year, while the MODIS LAI profile maintains lower LAI values. Comparatively speaking, the GLASS LAI values are the closest to the ground-based measurements. The MODIS data are very noisy, while all other LAI profiles are relatively smooth. Similar phenomena can also be observed in the other figures below. This is mainly due to the high sensitivity of the MODIS algorithm to surface reflectance uncertainties, especially at large LAI (Shabanov et al. 2005).

Figure 2.10 shows the temporal LAI trajectories for 2001–2002 at the Fundulea site with grass and cereal crop biome types according to the MODIS land cover map. The Fundulea site presents more interannual variability than other sites because its crops and cereals change from year to year. Although a good

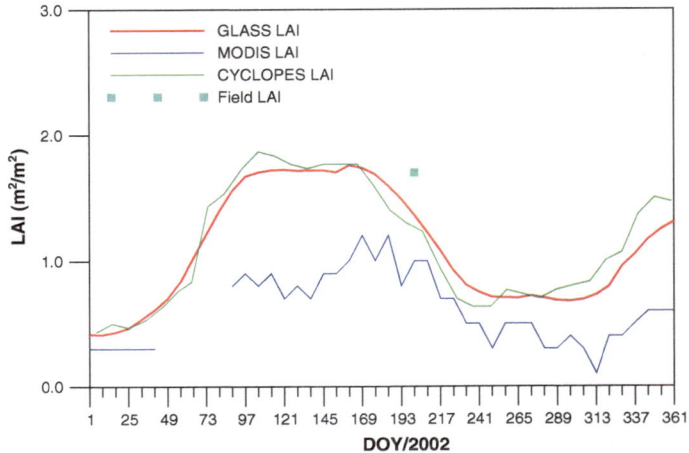

Fig. 2.9 Temporal profiles of GLASS, CYCLOPES, and MODIS LAI values for the Alpilles site for 2002. From Xiao et al. (2013), Copyright © 2003 reproduced with permission of IEEE

seasonality agreement is achieved among the GLASS, CYCLOPES, and MODIS LAI values, they are all underestimates in comparison with the BELMANIP mean LAI data at this site in 2001.

Figure 2.11 shows the temporal LAI trajectories for the Larose and Nezer sites with the conifer forest biome type according to the MODIS land cover map. The LAI products generally exhibit similar temporal trajectories, although with differences in magnitude. The strongest agreement is achieved between the GLASS and MODIS LAI values, aside from some fluctuations in the MODIS LAI values, while the CYCLOPES LAI values are significant underestimates throughout the entire growing season. The GLASS LAI values are in very good agreement with the BELMANIP mean LAI data at both sites.

As for broadleaf forests, the LAI temporal profiles at the Puechabon and Counami sites are illustrated in Fig. 2.12. The Puechabon site is largely dominated by trees of the species *Quercus ilex*, and is classified as an evergreen broadleaf forest according to the MODIS land cover map. Large discrepancies between products can be observed in the LAI magnitudes for 2001. The CYCLOPES LAI values exhibit an almost flat profile throughout the entire year. The GLASS and MODIS LAI values exhibit similar temporal trajectories, but with substantial differences in magnitude. The MODIS LAI values, except for dramatic fluctuations, are much larger than either the GLASS LAI values or the mean BELMANIP LAI values. However, the GLASS LAI values follow a smooth trajectory, and the GLASS LAI values outperform the other LAI products in terms of accuracy compared to the mean BELMANIP LAI data at this site. At the Counami site, visual inspection of the temporal profiles confirmed that the MODIS product values are extremely dubious, and most of the CYCLOPES LAI estimates are missing; in fact, there are only five CYCLOPES LAI values for the entire year.

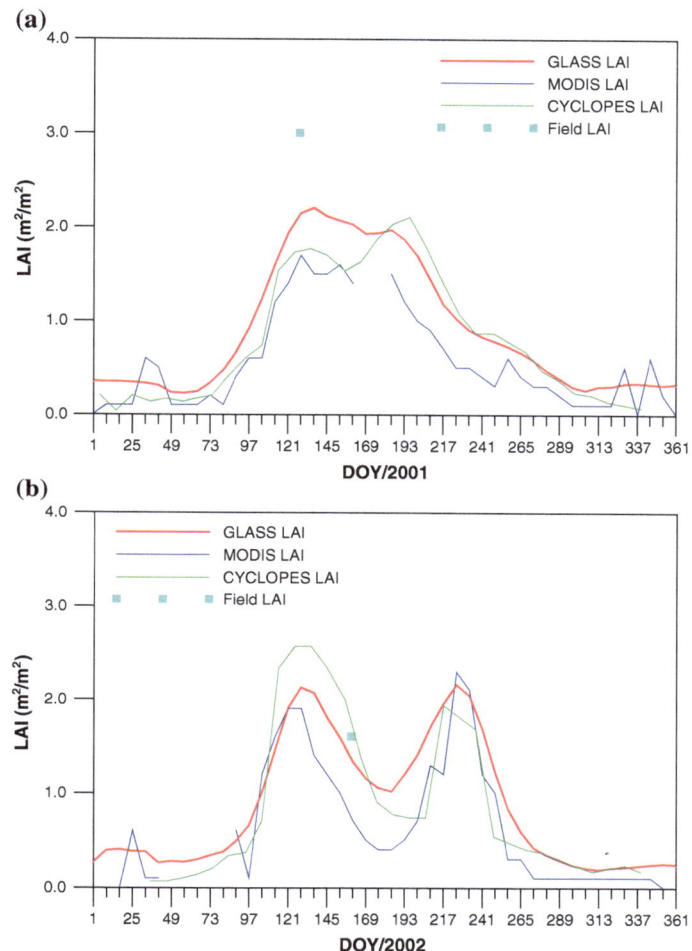

Fig. 2.10 Temporal profiles of GLASS, CYCLOPES, and MODIS LAI values for the Fundulea
site for 2001 (**a**) and 2002 (**b**). (Xiao et al. 2013), Copyright © 2003 reproduced with permission
of IEEE

Meanwhile, the GLASS LAI values are relatively smooth and close to the mean
BELMANIP LAI values.

The temporal profiles of the GLASS, MODIS, and CYCLOPES LAI values at
the Laprida and Larzac sites, which are of the savannah biome type, are provided
in Fig. 2.13. At the Laprida site, the LAI temporal profiles were generally in very
good agreement at the beginning of the growing season (Fig. 2.13a). However,
during the peak of the growing season, the MODIS LAI values became underestimates and the CYCLOPES LAI values became overestimates in comparison
with the mean BELMANIP LAI data at this site, while excellent agreement was

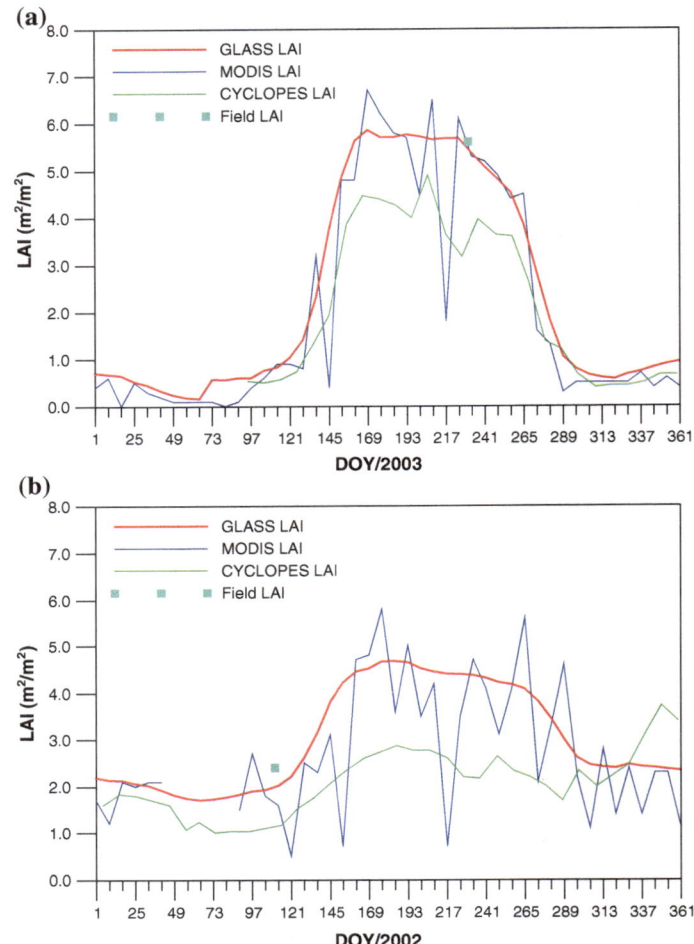

Fig. 2.11 Temporal profiles of GLASS, CYCLOPES, and MODIS LAI values at (**a**) the Larose site in 2003 and (**b**) the Nezer site in 2002. (Xiao et al. 2013), Copyright © 2003 reproduced with permission of IEEE

achieved between the GLASS LAI values and the ground measurements. At the Larzac site (Fig. 2.13b), the GLASS, MODIS, and CYCLOPES LAI values showed similar temporal trajectories, although with differences in magnitude. The strongest agreement was achieved between the GLASS and MODIS LAI values, although the MODIS LAI profile showed dramatic fluctuations; CYCLOPES maintained lower LAI values throughout the year. Figure 2.13b also shows that the GLASS, MODIS, and CYCLOPES LAI values were all overestimates when compared to the ground-based LAI values.

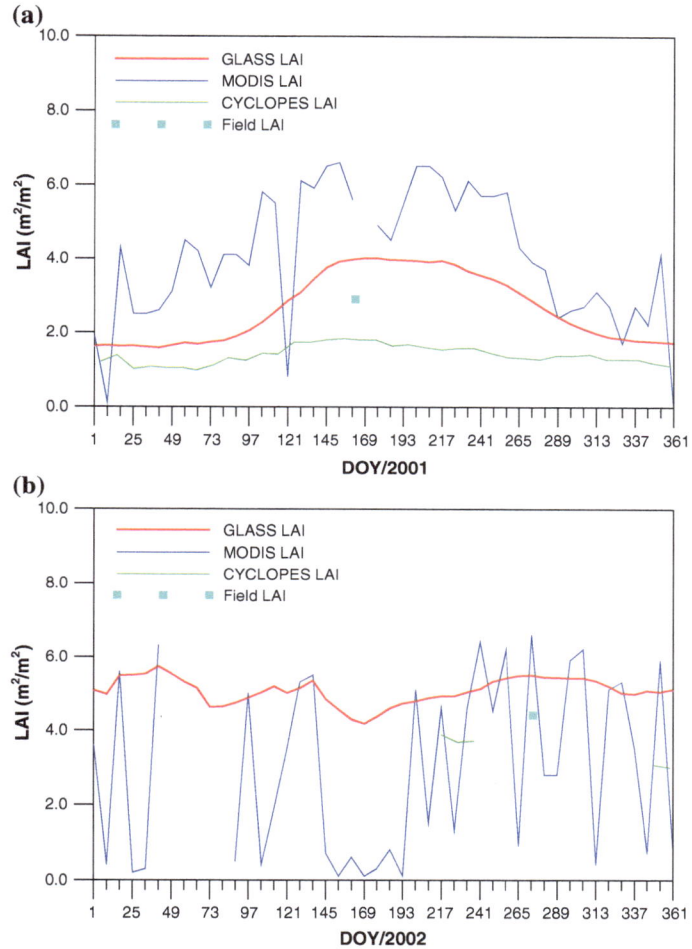

Fig. 2.12 Temporal profiles of GLASS, CYCLOPES, and MODIS LAI values at the Puechabon site for 2001 (**a**) and the Counami site for 2002 (**b**). (Xiao et al. 2013), Copyright © 2003 reproduced with permission of IEEE

To analyze the temporal consistency of the GLASS, MODIS, and CYCLOPES LAI products, the General Cartographic Transformation Package (GCTP) map projections library (U.S. Geological Survey 1993) was used to reproject the GLASS LAI values derived from MODIS surface reflectance and the MODIS and CYCLOPES LAI values into the geographic Lat/Lon projection, used in the GLASS LAI derived from AVHRR surface reflectance, and to resample them to exhibit a 0.05° spatial resolution using a spatial average sampling technique. To reduce potential co-registration errors, the average of all pixels in a 3 × 3 window, centered on the study sites, was used to compare the LAI time series. Figure 2.14

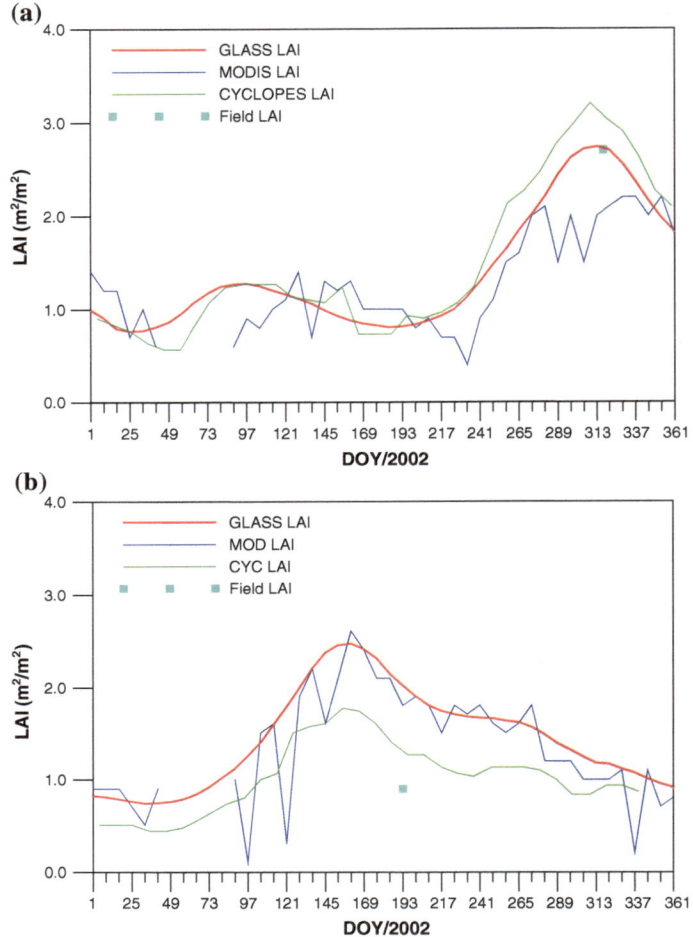

Fig. 2.13 Temporal profiles of GLASS, CYCLOPES, and MODIS LAI values at the Laprida site for 2002 (**a**) and the Larzac site for 2002 (**b**). (Xiao et al. 2013), Copyright © 2003 reproduced with permission of IEEE

shows the GLASS, CYCLOPES, and MODIS LAI profiles from 1982 to 2012 over the Bondville and NOBS—BOREAS sites.

The Bondville site (40.01°N, 88.29°W) is an agricultural site in the mid-western United States near Champaign, Illinois. The fields at this site were continuous no-till, with alternating years of soybean and maize crops (Meyers and Hollinger 2004). The NOBS—BOREAS site is located in the BOREAS Northern Study Area (NSA) at 55.89°N, 98.48°W and is surrounded by a black spruce-dominated forest of varying statures. The plots in Fig. 2.14 show that the GLASS LAI values retrieved from AVHRR and MODIS reflectance exhibit good time consistency,

(a)

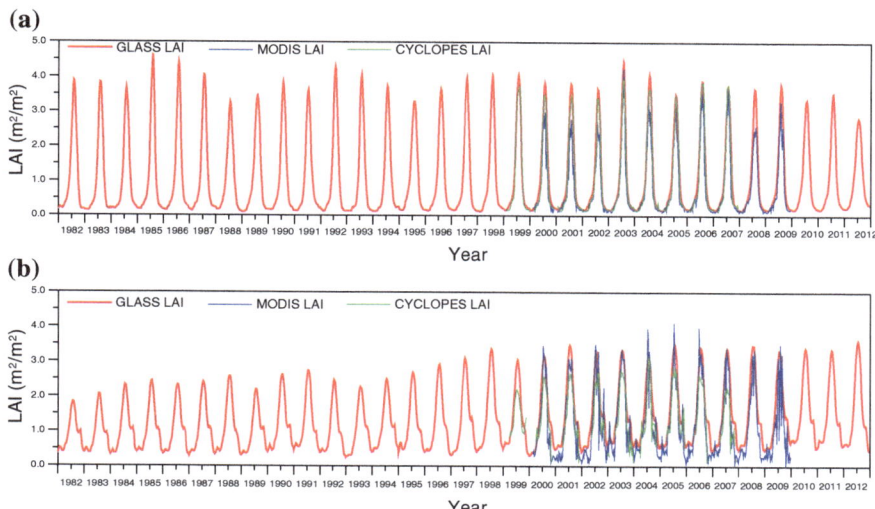

(b)

Fig. 2.14 Time profiles of LAI values from the GLASS, CYCLOPES and MODIS products at the Bondville (**a**) and NOBS–BOREAS (**b**) sites

and that all three LAI products are able to capture the seasonal change properties of vegetation to a similar degree, although the LAI peak value during the growing season may be different. Moreover, the GLASS and CYCLOPES LAI profiles are relatively smoother, while the MODIS LAI profiles fluctuate especially during the growing season.

In a separate study, Fang et al. (2013) compared five major global LAI products (MODIS, GEOV1, GLASS, GLOBMAP, and JRC-TIP) by examining their climatological and uncertainty information for different biome types. The results demonstrated that MODIS, GEOV1, GLASS, and GLOBMAP are generally consistent and show strong linear relationships between the products ($R^2 > 0.74$), with typical deviations of <0.5 for non-forest and <1.0 for forest biomes, and that the GEOV1 and GLASS products showed improvement over the earlier CYCLOPES and GLOBCARBON products in terms of spatial and temporal consistency and similarity to MODIS (Fang et al. 2013).

2.4.2 Direct Validation

The Validation of Land European Remote Sensing Instruments (VALERI) project has developed a globally distributed network of sites and a standard method to measure biogeophysical variables of interest directly at the proper spatial and temporal scales. This network contains 33 sites and 52 LAI reference maps. The LAI reference maps were derived by determining the transfer function between the

reflectance values of high-spatial-resolution SPOT images and LAI ground measurements. The SPOT images were acquired by HRVIR1 on SPOT4 during or just
before or after the ground campaign using the UTM projection. The ground
measurements consisted mainly of gap-fraction measurements obtained using
LAI-2000 measurements or hemispherical photographs. For each Elementary
Sampling Unit (ESU), the hemispherical images were processed using the CAN-
EYE software (Version 3.6) developed at INRA-CSE. Among the 52 LAI reference maps, only 22 LAI reference maps provide true LAI values from 2000 to
2004. These true LAI maps were aggregated to produce a moderate resolution and
used to compare and evaluate the GLASS LAI products. Characteristics of the
validation sites are shown in Table 2.4.

The GLASS, MODIS, and CYCLOPES LAI products were compared directly
with the same set of LAI reference maps. Figure 2.15 presents the accuracy of
each product as quantified by the regression function, R-squared, and RMSE. Each

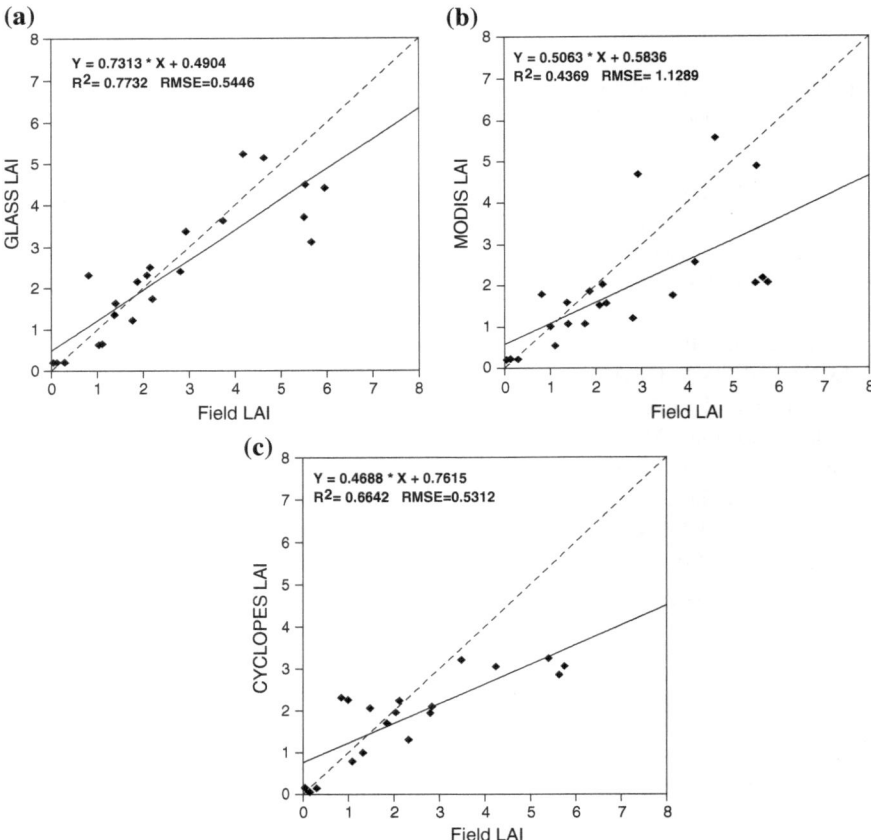

Fig. 2.15 GLASS (**a**), MODIS (**b**), and CYCLOPES (**c**) LAI products versus LAI reference map
scatterplots for direct validation

scatter diagram has 22 plots corresponding to the 22 LAI maps, except for
CYCLOPES. Because there are two fill values of the CYCLOPES LAI corresponding to two of the LAI reference maps, only 20 plots are shown in Fig. 2.15c.
The best agreement was achieved by the GLASS LAI ($R^2 = 0.77$, RMSE = 0.54),
followed by the CYCLOPES LAI ($R^2 = 0.66$, RMSE = 0.53), and the MODIS
LAI ($R^2 = 0.44$, RMSE = 1.13). It can be concluded that the GLASS LAI has
better consistency with ground measurement data than the other two products.

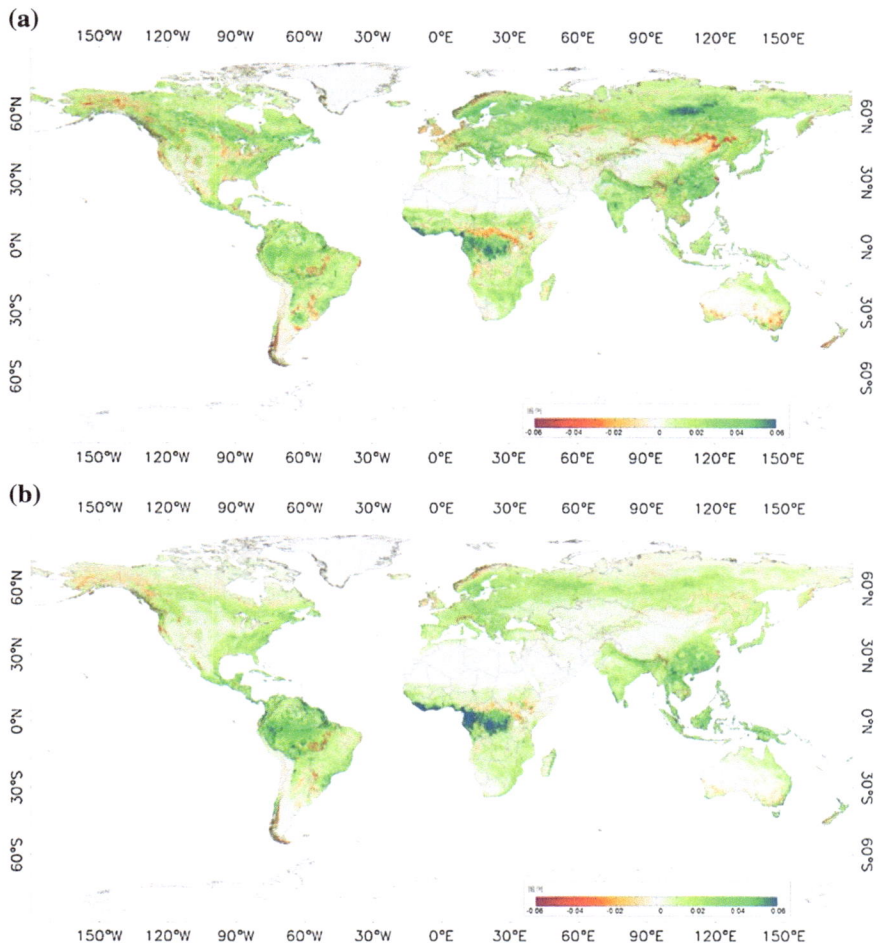

Fig. 2.16 Linear trend based upon annual maximum LAI (**a**) and annual average LAI (**b**) for
1982–2011

2.5 Preliminary Analysis and Applications

2.5.1 Spatial and Temporal Variation of Global LAI

The GLASS LAI product for 1982–2011 was used to analyze the global temporal and spatial variations of vegetation for periods of 30 years. For each pixel, a least-squares fit was applied to the annual maximum LAI and annual average LAI from 1982 through 2011, to determine the slope coefficients of the linear regressions that represent annual trends (shown in Fig. 2.16).

The linear trends based upon the annual maximum LAI (Fig. 2.16a) shows a significantly decreasing trend of vegetation growth on the Europe's Atlantic coast, in Alaska, in the northeastern part of the Mongolian Plateau, and along the northern margin of the African rainforest. On the other hand, a significantly increasing trend of vegetation growth was found in Central Siberia, the Gulf of Guinea along the west coast of Africa, and the Congo Basin.

The linear trends based on the annual average LAI (Fig. 2.16b) shows a significantly increasing trend of vegetation growth in the African countries along the Gulf of Guinea, such as Liberia, Ivory Coast, Ghana, Cameroon, Equatorial Guinea, Gabon, Congo-Kinshasa, and Congo-Brazzaville, and a significantly decreasing trend of vegetation growth in northern Mongolia, eastern Inner Mongolia in China, the Central African Republic, Central Nigeria, northern Cameroon, southern South Sudan, northern Uganda, southern Ethiopia, Alaska, western Canada, southeastern and western Paraguay, northern Chile, northern Argentina, the southeastern margin of the Amazon plains, and south-central, and southwestern Australia.

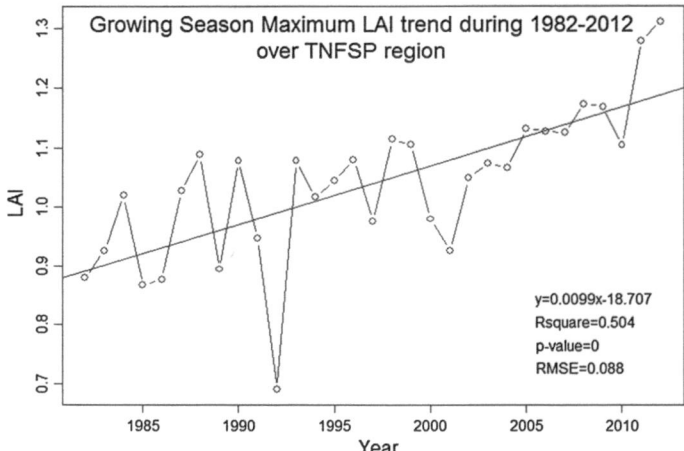

Fig. 2.17 Average values of annual maximum LAI over the three-north China area during 1982–2012

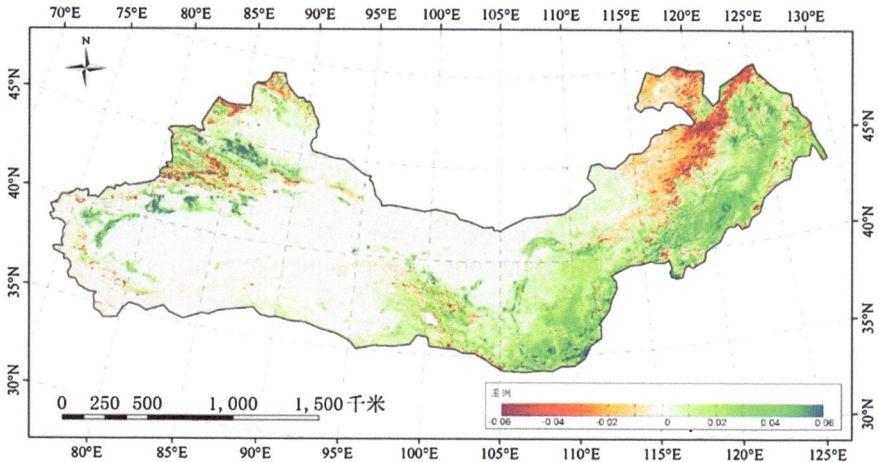

Fig. 2.18 Linear trend based on annual maximum LAI over the Three-North China area for 1982–2012

2.5.2 Three-North China Afforestation

The Three-North Forest Shelterbelt Program has been carried out in northeastern China, northern China, and northwestern China since 1978. Other ecological restoration programs have also been carried out in these regions. These programs have greatly improved the ecological status of the Three-North Forest Shelterbelt region, making it one of the most significant areas of positive vegetation change in China over the past 30 years. Figure 2.17 shows the average of the annual maximum LAI over the area during 1982–2012, and it is interesting to note an overall increasing trend.

The linear trend based on the annual maximum LAI over the Three-North Forest Shelterbelt region is demonstrated in Fig. 2.18. In addition to the eastern region of Inner Mongolia, all other Three-North project areas show a significantly increasing trend for 1982–2012 (Fig. 2.18).

2.6 Summary

The GLASS LAI product has been generated using GRNNs for 1981–2012 from time-series MODIS and AVHRR reflectance data. The GRNNs were trained using fused LAI values calculated from the MODIS and CYCLOPES LAI products and the corresponding reprocessed MODIS/AVHRR reflectance values. The input to the GRNNs was the time-series reprocessed MODIS/AVHRR reflectance values for an entire year, and the output of the GRNNs was the LAI profile for 1 year.

By computing the RMSE and R-squared values of each product over the LAI reference maps at various validation sites, it has been demonstrated that the accuracy of the GLASS LAI product is clearly better than that of MODIS and CYCLOPES. Moreover, the GLASS LAI is more temporally continuous and spatially complete than the other products tested. The spatial patterns generated by GLASS are reasonable and consistent with good quality MODIS and CYCLOPES LAI values. The GLASS and CYCLOPES LAI products have smoother trajectories compared to the erratic fluctuations of the MODIS LAI. The GLASS LAI has more realistic and reasonable trajectories representing seasonal variations, especially for forest.

References

Baret F, Hagolle O et al. (2007) LAI and fAPAR CYCLOPES global products derived from VEGETATION. Part 1: Principles of the algorithm. Remote Sensing of Environment, 110(3):275–286

Baret F, Morissette JT, Fernandes RA, Champeaux JL, Myneni RB, Chen J, Plummer S, Weiss M, Bacour C, Garrigues S, Nickeson JE (2006) Evaluation of the representativeness of networks of sites for the global validation and intercomparison of land biophysical products: proposition of the CEOS-BELMANIP. IEEE Trans Geosci Remote Sens 44:1794–1803

Buermann W, Dong J, Zeng X, Myneni RB, Dickinson RE (2001) Evaluation of the utility of satellite-based vegetation leaf area index data for climate simulations. J Clim 14:3536–3550

Chen JM (1996) Optically-based methods for measuring seasonal variation of leaf area index in boreal conifer stands. Agric For Meteorol 80:135–163

Chen JM, Menges CH, Leblanc SG (2005) Global mapping of foliage clumping index using multi-angular satellite data. Remote Sens Environ 97:447–457

Chen J, Deng F, Chen M (2006) Locally adjusted cubic-spline capping for reconstructing seasonal trajectories of a satellite-derived surface parameter. IEEE Trans Geosci Remote Sens 44:2230–2238

Duan Q, Sorooshian S, Gupta V (1992) Effective and efficient global optimization for conceptual rainfall-runoff models. Water Resour Res 28:1015–1031

Fang H, Jiang C, Li W, Wei S, Baret F, Chen JM, Garcia-Haro J, Liang S, Liu R, Myneni RB, Pinty B, Xiao Z, Zhu Z (2013) Characterization and intercomparison of global moderate resolution leaf area index (LAI) products: analysis of climatologies and theoretical uncertainties. J Geophys Res Biogeosci 118. doi:10.1002/jgrg.20051

Hansen JV, Meservy RD (1996) Learning experiments with genetic optimization of a generalized regression neural network. Decis Support Syst 18:317–325

Knyazikhin Y, Martonchik JV, Diner DJ, Myneni RB, Verstraete M, Pinty B, Gobron N (1998) Estimation of vegetation canopy leaf area index and fraction of absorbed photosynthetically active radiation from atmosphere-corrected MISR data. J Geophys Res Atmos 103:32239–32256

Leung MT, Chen A-S, Daouk H (2000) Forecasting exchange rates using general regression neural networks. Comput Oper Res 27:1093–1110

Liang S, Li X, Wang J (2012) (ed) Advanced remote sensing: terrestrial information extraction and applications. Academic Press, Oxford

Liang SL (2007) Recent developments in estimating land surface biogeophysical variables from optical remote sensing. Prog Phys Geogr 31:501–516

Meyers TP, Hollinger SE (2004) An assessment of storage terms in the surface energy balance of maize and soybean. Agric For Meteorol 125:105–115

Myneni RB, Hoffman S, Knyazikhin Y, Privette JL, Glassy J, Tian Y, Wang Y, Song X, Zhang Y, Smith GR, Lotsch A, Friedl M, Morisette JT, Votava P, Nemani RR, Running SW (2002) Global products of vegetation leaf area and absorbed PAR from year one of MODIS data. Remote Sens Environ 83:214–231

Myneni RB, Nemani RR, Shabanov NV, Knyazikhin Y, Morisette JT, Privette JL, Running SW (2007) LAI and FPAR. In: NASA Earth System Data Records (ESDR) White Papers

Pisek J, Chen JM, Lacaze R, Sonnentag O, Alikas K (2010) Expanding global mapping of the foliage clumping index with multi-angular POLDER three measurements: evaluation and topographic compensation. ISPRS J Photogramm Remote Sens 65:341–346

Shabanov NV, Dong H, Wenze Y, Tan B, Knyazikhin Y, Myneni RB, Ahl DE, Gower ST, Huete AR, Aragao LEOC, Shimabukuro YE (2005) Analysis and optimization of the MODIS leaf area index algorithm retrievals over broadleaf forests. IEEE Trans Geosci Remote Sens 43:1855–1865

Specht DF (1991) A general regression neural network. IEEE Trans Neural Netw 2:568–576

Tang H, Yu K, Geng X, Zhao Y, Jiang K (2013) A time series method for cloud detection applied to MODIS surface reflectance images. Int J Digit Earth. doi:10.1080/17538947.2013.833313

U.S. Geological Survey N.M.D. (1993) GCTP General Cartographic Transformation Package Software Documentation

Weiss M, Baret F, Garrigues S, Lacaze R (2007) LAI and fAPAR CYCLOPES global products derived from VEGETATION. Part 2: validation and comparison with MODIS collection 4 products. Remote Sens Environ 110:317–331

Xiang Y, Xiao Z, Liang S, Wang J, Song J (2013) Validation of global land surface satellite (GLASS) leaf area index product derived from time series MODIS reflectances. J Remote Sens (to be published)

Xiao Z, Liang S, Wang J, Song J, Wu X (2009) A Temporally integrated inversion method for estimating leaf area index from MODIS data. IEEE Trans Geosci Remote Sens 47:2536–2545

Xiao Z, Liang S, Wang J, Chen P, Yin X, Zhang L, Song J (2013) Use of general regression neural networks for generating the GLASS leaf area index product from time-series MODIS surface reflectance. IEEE Trans Geosci Remote Sens. doi:10.1109/TGRS.2013.2237780

Xiao Z, Liang S, Wang J, Jiang B, Li X (2011) Real-time retrieval of leaf area index from MODIS time series data. Remote Sens Environ 115:97–106

Xiao Z, Wang J, Liang S, Zhou H, Li X, Zhang L, Jiao Z, Liu Y, Fu Z (2012) Variational retrieval of leaf area index from MODIS time series data: examples from the Heihe river basin, northwest China. Int J Remote Sens 33:730–745

Chapter 3
Shortwave Albedo

Abstract This chapter describes the algorithm, analysis, preliminary validation, and application of the GLASS albedo product. Unlike traditional remote sensing products, the GLASS albedo product was generated in two steps: the first step retrieved albedo values from remote sensing data using two direct-estimation algorithms, and the second step applied a statistics-based temporal filter to the directly estimated albedo values to generate a high-quality, gapless final product. The GLASS albedo product has been validated using FLUXNET observation data and compared with the MODIS instrument Bidirectional Reflectance Distribution Function (BRDF)/albedo product. The results show the high quality and accuracy of the GLASS albedo product and its suitability for long-term global environmental change studies. It is one of the longest duration (1981–2010) satellite shortwave albedo products in the world.

Keywords Albedo · Shortwave radiation · Angular bin · Temporal filter · MODIS · AVHRR · GLASS

3.1 Background

Land surface albedo is one of the major driving factors in the climate system because it determines how much solar shortwave radiation will be absorbed at the land surface (Dickinson 1983; Liang 2004; Liang et al. 2010, 2013a; Mason 2005). The variation of land surface albedo is hard to predict because it is affected by many factors (Gao et al. 2005) such as snow, rain, and vegetation status. Satellite remote sensing is the most effective technique for mapping the spatial and temporal distribution of global land surface albedo.

Many global or regional land surface albedo products have been made available. They are derived either from polar-orbiting satellite data, e.g., MODIS (Gao et al. 2005; Lucht et al. 2000; Schaaf et al. 2002), Polarization and Directionality

S. Liang et al., *Global LAnd Surface Satellite (GLASS) Products*,
SpringerBriefs in Earth Sciences, DOI: 10.1007/978-3-319-02588-9_3,
© The Author(s) 2014

of the Earth's Reflectances (POLDER) (Bacour and Breon 2005; Leroy et al. 1997; Maignan et al. 2004), MERIS (Muller et al. 2007), CERES (Rutan et al. 2006), MISR (Diner 2008; Weiss et al. 1999), and VEGETATION (Geiger and Samain 2004), or from geostationary satellite data, e.g., MFG (Govaerts et al. 2008; Pinty et al. 2000) and MSG (Geiger et al. 2005; van Leeuwen and Roujean 2002). The MODIS MCD43 BRDF (Bidirectional Reflectance Distribution Function) and albedo products, generated by NASA's MODLAND team, provide a global albedo map in a continuous time series from 2000 to the present (Schaaf et al. 2008). The MCD43 products are the most complete and widely recognized dataset for global albedo, and their algorithm has had a profound influence on other albedo inversion algorithms. Before 2000, however, the available global albedo datasets are very limited. The published AVHRR albedo products (Hu et al. 2000; Saunders 1990; Strugnell et al. 2001; Trishchenko et al. 2008), the POLDER albedo product, and the VEGETATION albedo product cannot compose a continuous time series.

Global environmental change research demands a spatially intact, long-term continuous global albedo product. However, satellite albedo products from a single sensor are limited by sensor characteristics and acquisition conditions as well as cloud coverage. They usually contain gaps that may cause significant troubles for data users. The preferred solution is to combine data from multiple sensors to derive one complete and consistent dataset. This is why the GLASS shortwave albedo product has been produced.

3.2 Algorithms

3.2.1 Algorithm Overview

To provide a high-quality, user-friendly data product, the GLASS albedo product is produced in two steps. In the first step, the albedo values are retrieved from satellite data by inversion of the radiative transfer model. The results from the first step are called the intermediate product. In the second step, different kinds of intermediate products are merged and filtered to generate a unique and seamless final product using statistics and a priori knowledge.

The GLASS albedo product after Feburary 24, 2000 has a spatial resolution of 1 km and a temporal rolling frequency of one day, and its data source is MODIS data from both Terra and Aqua satellites. Two retrieval algorithms were used to derive intermediate albedo products. They are both regression algorithms improved with the angular bin (AB) method first proposed by Liang et al. (1999, 2005) and further developed by Qu et al. (2013).

The angular bin 1 (AB1) algorithm builds a linear regression equation between ground surface directional reflectance and broadband albedo, specifically short-wave white-sky-albedo (WSA) and black-sky-albedo (BSA) corresponding to the solar angle at local noon. The input of AB1 is the re-processed MOD09GA/

MYD09GA product released by NASA (Vermote et al. 2002). The output of AB1 is either the intermediate product GLASS02A21 corresponding to Terra, or GLASS02A22 corresponding to Aqua. Both products have a spatial resolution of 1 km and a temporal resolution of one day.

The angular bin 2 (AB2) algorithm is similar to the AB1 algorithm, but builds a linear regression equation between top-of-atmosphere (TOA) directional reflectance and broadband albedo. The input of AB2 is the MOD021 km/MYD021 km product released by NASA. The output of AB2 is either the intermediate product GLASS02A23 corresponding to Terra, or GLASS02A24 corresponding to Aqua. Both products also have a spatial resolution of 1 km and a temporal resolution of one day.

Therefore, four intermediate products after 2000, corresponding to different algorithms and data sources, are available. Instead of sending them all together to the user, it was decided to merge them into a final product, called GLASS02A06. The module that does this is called Statistics-based Temporal Filtering (STF) (Liu et al. 2013a). Its basic idea is to merge and filter intermediate products in the temporal dimension based on the Bayes principle. Gaps and flaws are also accounted for during this process.

The GLASS albedo product before 2000 is based on AVHRR data with a spatial resolution of 5 km and a temporal rolling frequency of 8 days. The Long-term Land Data Record (LTDR) project has archived AVHRR data from 1981 to 2000 (Pedelty et al. 2007). Its version 3 released AVHRR data have been geometrically and atmospherically corrected. The GLASS production system combines the AB1 and STF algorithms in a pipeline program to generate the albedo product, called GLASS02A05, from the AVHRR dataset of LTDR. The GLASS albedo product before 2000 has a spatial resolution of $0.05°$ and a temporal resolution of eight days.

Figure 3.1 presents the flowchart of the albedo production line in the GLASS production system (Liu et al. 2013b). The production process is composed of two steps. Step 1 interprets the satellite data and derives the intermediate products. Step 2 merges the intermediate products and corrects any flaws.

3.2.2 The AB1 Algorithm

Most satellite albedo estimation algorithms consist of three procedures (Liang et al. 2010, 2012; Liu et al. 2012) atmospheric correction, surface directional reflectance modeling, and narrowband-to-broadband conversion. However, the errors from each procedure may accumulate and affect the final accuracy of the albedo product. An alternative approach, the so-called "direct estimation algorithm", can be used to estimate surface albedo directly from TOA observations. This approach combines all procedures in one step through regression analysis with the sole aim of creating a best-estimate broadband albedo. In an earlier study, Liang et al. (1999) developed such a direct retrieval algorithm using a

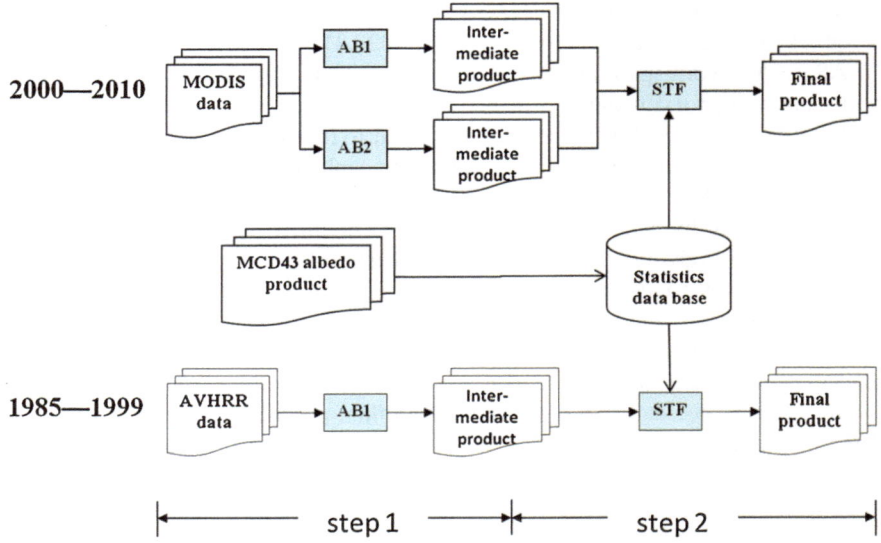

Fig. 3.1 Flowchart of the albedo product generation in the GLASS production system (Liu et al. 2013b)

feed-forward neural-network. It was later improved by using linear regression analysis in each ABs and applied to MODIS data (Liang 2003). Further improvements produced highly accurate daily snow and ice albedo values more efficiently with a mean bias of less than 0.02 and residual standard error of 0.04 (Liang et al. 2005). If a surface reflectance product (after atmospheric correction of the TOA observations) is available, the last two steps can be combined to convert the directional reflectance to broadband albedo.

A multivariate linear regression relationship is assumed between the surface broadband albedo and the surface bidirectional reflectances (BDR) in the MODIS visible and near infrared spectral bands, which can be expressed by the following equation:

$$A = c_0(\theta_i, \, \theta_r, \, \varphi) + \sum_{i=1}^{n} c_i(\theta_i, \, \theta_r, \, \varphi)\rho_i(\theta_i, \, \theta_r, \, \varphi) \tag{3.1}$$

where A represents the surface albedo (more specifically, the shortwave (0.3–5 µm) WSA or BSA at a certain solar zenith angle); c_i are the regression coefficients; and ρ_i are the surface directional reflectances in band i, which are all functions of solar/view zenith angles.

The AB1 algorithm is used to build the linear regression relationship between the surface BDR on the MODIS bands and the broadband albedo for each grid element. It has the advantages of simple computation, low input data requirements, and full consideration of the surface bidirectional and spectral characteristics.

The first step in the surface albedo inversion is the calculation of the regression coefficients, c_i. Because of the BDR characteristic of the natural land surface, the regression coefficients c_i change with the solar/view zenith angles. For convenience in numerical calculation, the solar/view geometry space was divided into a three-dimensional grid defined by the solar zenith, the view zenith, and the relative azimuth, in which each grid element is called an AB. Reducing the grid size would improve precision but increase use of computer resources; therefore, a balance must be sought between these considerations. The solar zenith angle of the centers of the ABs varies from 0 to 80° in 4° increments (i.e., 0, 4,..., 80°). The view zenith angle of the centers of the ABs varies from 0 to 64° in 4° increments (i.e., 0, 4,..., 64°). The relative azimuth angle of the centers of the ABs varies from 0 to 180° in 20° increments (i.e., 0, 20,..., 180°).

The regression coefficients of the AB1 algorithm are obtained using training data; therefore, a training dataset containing the multi-band BDR and the broad-band albedo is needed. The quality and representativeness of this training dataset strongly influences the performance of the AB1 algorithm. Similar to the method proposed by Cui et al. (2009), a training dataset that presents various global surfaces was built based on the POLDER-BRDF dataset. Although the POLDER-BRDF observations cover many different observation angles, it was impossible to ensure enough observations within each AB. Consequently, the POLDER-BRDF database must be interpolated and extrapolated to build the complete training dataset. It can then be integrated to obtain the broadband albedo.

3.2.2.1 Fitting and Interpolation Method for the POLDER-BRDF Database

The Polarization and Anisotropy of Reflectances for Atmospheric Sciences coupled with Observations from a Lidar (PARASOL), launched by the Centre National d'Etudes Spatiales (CNES) on October 18th, 2004, hosted the third-generation POLDER, which can collect global polarization and directional data for reflected solar radiation from the atmosphere and the Earth's surface. The spatial resolution of the POLDER product is 6×7 km, providing abundant angular, spectral, and polarization information. These data are the latest currently available multi-angle satellite remote sensing data and provide much information about the Earth surface, atmospheric aerosols, and clouds. The imaging characteristics of the POLDER sensor enable it to obtain observation data for as many as 14 angles, up to 60°, in every track. Each sample of the POLDER-BRDF dataset is composited using monthly observed data. The data processing algorithm fully considers the impacts of cloud detection, molecule absorption correction, and aerosols in the stratosphere and troposphere.

When the solar zenith angle is greater than 60° or less than 20°, observations are particularly rare, and therefore the POLDER-3/PARASOL database must be interpolated and extrapolated to build the complete training dataset. The commonly used method for fitting and predicting BRDF is the linear kernel-driven model (Roujean et al. 1992; Wanner et al. 1995), in which the Bidirectional Reflectance Factor (BRF) is expressed as follows:

$$R(\theta_s, \theta_v, \varphi; \lambda) = f_{\text{iso}}(\lambda) + f_{\text{vol}}(\lambda)k_{\text{vol}}(\theta_s, \theta_v, \varphi) + f_{\text{geo}}(\lambda)k_{\text{geo}}(\theta_s, \theta_v, \varphi) \qquad (3.2)$$

where θ_s is the solar zenith angle; θ_v is the view zenith angle; φ is the relative azimuth angle; λ is the wavelength; k_{vol}, k_{geo} are the volume scattering and geo-optical kernels respectively; and f_{iso}, f_{vol}, f_{geo} are the coefficients of the isotropic, volume scattering, and geo-optical kernels, respectively.

The original linear kernel-driven model proposed by Wanner et al. (1995) is particularly suitable for accounting for the backward-scattering effect of directional reflectance from the vegetation canopy. Because of the strong forward-scattering effects of directional reflectances of snow and ice in polar areas, a forward-scattering kernel is needed. After extensive comparisons of several different models, the original linear kernel-driven model was revised as follows:

$$R(\theta_s, \theta_v, \varphi; \lambda) = f_{\text{iso}}(\lambda) + f_{\text{geo}}(\lambda)k_{\text{geo}}(\theta_s, \theta_v, \varphi) + f_{\text{vol}}(\lambda)k_{\text{vol}}(\theta_s, \theta_v, \varphi) + f_{\text{fwd}}(\lambda)k_{\text{fwd}}(\theta_s, \theta_v, \varphi)$$
$$(3.3)$$

where $k_{\text{fwd}}(\theta_s, \theta_v, \varphi)$ and $f_{\text{fwd}}(\lambda)$ are the forward-scattering kernel and its coefficients, respectively. A revised Ross-Thick kernel, which accounts for the hot-spot effect (Maignan et al. 2004) was used as the volume scattering kernel, and the Li-Sparse-R kernel (Strahler et al. 1999) was used as the geometric optical kernel.

The forward-scattering kernel was added to describe the forward-scattering effects for snow surfaces and was derived by simplifying the Rahman-Pinty-Verstraete (RPV) model (Rahman et al. 1993) and setting the parameters to typical values.

$$k_{\text{fwd}}(\theta_s, \theta_v, \varphi) = \frac{\cos^{k-1}\theta_s \cos^{k-1}\theta_v}{(\cos\theta_s + \cos\theta_v)^{1-k}} \cdot \frac{1-g^2}{(1+g^2-2g\cos(\pi-\xi))^{3/2}} - \frac{1+g}{2^{1-k}(1-g)^2}$$
$$(3.4)$$

where $g = 0.0667$ and $k = 0.846$.

3.2.2.2 Band Conversions from POLDER to MODIS

Because the training datasets are obtained by POLDER, but the inversion algorithm is applied to MODIS data, the relationship between the surface reflectance observations of the two sensors in each band must be established and applied to the band conversions. The principle of band conversion relies on the premise that the surface reflectances of typical land surface in different bands are correlated with each other. This premise has been widely adopted in Earth observation studies. A continuous spectrum of typical surface features is collected first, and then the corresponding reflectance is calculated through the spectral response functions in each POLDER and MODIS band. Finally, the linear conversion relationship is developed according to the statistical properties of the data in these bands.

A set of 493 field measured spectra of typical land covers, including crops, natural vegetation, ground, sand, water, snow, and ice, were used to derive the band conversion coefficients, which are listed in Table 3.1.

The RMSE values of the first four MODIS bands are smaller than those of the latter three bands; therefore, a greater uncertainty was added to the band conversion of the latter three bands, which was considered when the albedo regression equations were built.

The kernel-driven model was also used in the calculation of the surface albedo corresponding to each dataset; here, the broadband albedo in the 0.3–5 µm band range was represented by WSA and BSA. Because the black-sky albedo is a function of the solar zenith angle, all black-sky albedos between 0 and 80° were calculated at 5° intervals. As a feature of the kernel-driven model, the narrowband albedo is the weighted sum of the integration of the kernel functions, with the kernel coefficients as weights. The black-sky and white-sky albedos can be calculated as follows (Lucht et al. 2002):

$$\alpha_{bs}(\theta_s, \lambda) = \sum_k f_k(\lambda) h_k(\theta_s) \tag{3.5}$$

$$\alpha_{ws}(\lambda) = \sum_k f_k(\lambda) H_k \tag{3.6}$$

where $h_k(\theta_s)$ is the integral of the linear kernel-driven model kernels k when the solar zenith angle is θ_s, and H_k is the integral of $h_k(\theta_s)$. f_k are the coefficients of the linear kernel-driven model kernels k.

The narrowband-to-broadband conversions can be expressed as (Liang 2001):

$$\alpha = c_0 + \sum_{i=1}^{n} c_i \alpha(\lambda_i) \tag{3.7}$$

where α is the broadband albedo, $\alpha(\lambda)$ is the spectral (narrowband) albedo, and c_i ($i = 0, 1, \cdots, n$, where n is the number of bands) are the coefficients for band conversions. The conversion coefficients (Table 3.2) for narrowband-to-broadband conversions have been derived in several previous studies (Liang 2001; Stroeve et al. 2005).

3.2.2.3 Use of Land Cover Types

Bidirectional reflectance characteristics vary by land cover type. Although the AB1 surface albedo inversion algorithm is based on regression, which can adapt to variation in BDR by dividing the solar/view geometry space into a three-dimensional grid, the linear regression model contains approximation error.

Table 3.1 Band conversion coefficients from POLDER to MODIS

Band name	MODIS-b1-648	MODIS-b2-859	MODIS-b3-466	MODIS-b4-554	MODIS-b5-1244	MODIS-b6-1631	MODIS-b7-2119
POLDER-b1-490	0.024593	0.032882	0.912575	0.207528	-0.356931	-1.039259	-1.153855
POLDER-b2-565	0.306280	-0.036396	0.143223	0.613727	-0.015207	-0.441590	-0.504465
POLDER-b3-670	0.691690	-0.011160	-0.060149	0.121219	0.418777	1.540050	1.822514
POLDER-b4-765	-0.044711	0.299618	0.005727	0.119292	-0.117775	-0.186372	-0.114003
POLDER-b5-865	-0.005402	0.645341	0.021614	-0.026334	-0.580513	-0.792087	-0.776048
POLDER-b6-1020	0.030160	0.081619	-0.026446	-0.043885	1.488419	1.267713	0.918089
Offset	0.004258	0.001039	-0.007456	-0.000374	0.022993	0.070931	0.070192
RMSE	0.003089	0.003553	0.004188	0.004286	0.020661	0.041961	0.047242

Table 3.2 Narrowband-to-broadband conversion coefficients

MODIS bands (mm)	c0 (offset)	c1 (0.62– 0.67)	c2 (0.84– 0.87)	c3 (0.46– 0.48)	c4 (0.54– 0.56)	c5 (1.23– 1.25)	c6 (1.63– 1.65)	c7 (2.11– 2.15)
Snow/ice free	−0.0015	0.1600	0.2910	0.2430	0.1160	0.1120	0.0000	0.0810
Snow/ice	−0.0093	0.1574	0.2789	0.3829	0.0000	0.1131	0.0000	0.0694

Therefore, it is necessary to introduce land cover type information and to subdivide the training samples into several classes to reduce the uncertainty of the linear regression model.

If global land use/land cover products (such as the MODIS Land Use/Land Cover Products) are used to support the albedo inversion, the inputs to the algorithm will increases and its applicability will decreases. Simultaneously, new problems must be considered: first, error exists in the global land cover products, especially error caused by mixed pixels in the 1 km-resolution data; second, surface albedo is a rapidly changing physical parameter, but the land cover types are determined according to the surface long-term cover state. For instance, the albedo of cropland changes significantly after snowfall but its land cover type is still cropland.

Therefore, in this study, a relatively simple classification method directly based on remote sensing observations was used. Basically, the POLDER-BRDF datasets were divided into three classes: vegetation, bare ground, and snow/ice. The following classification criteria were used: (1) for each observation, if the NDVI value was greater than 0.2, the pixel was categorized as vegetation; (2) if the reflectance of the blue or red channel was greater than 0.3, the pixel was categorized as ice/snow; (3) the remaining pixels were categorized as bare ground. In the AB regression procedure, the datasets for different classes were analyzed separately.

In the generation of the training dataset, the average reflectance and average NDVI of each spectral band formed the basis of this classification. To maintain the consistency of the calculated albedo during the transition from one class to another, intermediate classes were designated. For instance, pixels with a corresponding NDVI value between 0.18 and 0.24 would be categorized as "intermediate class A," and pixels with blue reflectance between 0.24 and 0.4 would be categorized as "intermediate class B." Other pixels were temporarily categorized as pure vegetation, pure bare ground, and pure ice/snow. Using this procedure, the POLDER-BRDF datasets were categorized into five classes, resulting in 4737 pure vegetation datasets, 2401 pure ground datasets, 627 pure ice/snow datasets, 1136 intermediate class A datasets, and 123 intermediate class B datasets. During the regression analysis, the categories were further aggregated into three classes: vegetation (pure vegetation + intermediate class A), ground (pure ground + intermediate class A + intermediate class B), and snow/ice (pure snow/ice + intermediate class B). In this way, proper overlapping guarantees that the

Fig. 3.2 Scatter diagram of the average reflectance in the blue band and the average NDVI of the POLDER-BRDF dataset. (*Blue, red, green, yellow,* and *purple points* represent pure snow/ice, pure ground, pure vegetation, intermediate class, A and intermediate class B, respectively) (Qu et al. 2013). Copyright © 2013 reproduced with permission of IEEE

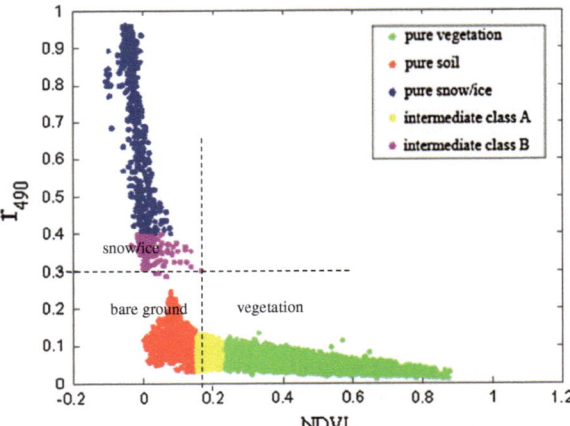

regression result has the desirable consistency in the transition regions from one class to another. Figure 3.2 shows a scatter diagram of the five categories of training data in the NDVI-r_{490} feature space, in which r_{490} represents the average reflectance in the blue band.

3.2.2.4 Regression Method

Equation (3.1) describes the relationship between the surface multi-band BDR factor and the broadband albedo, in which the regression coefficients are undetermined. For each solar/view AB, a group of regression coefficients, i.e., $c_i | i = 0, \dots, n$, needs to be estimated according to the training data.

To solve the equations using the least-squares method, the equations can be written in matrix form:

$$\mathbf{Y} = \mathbf{AX} \tag{3.8}$$

where **X** is the matrix containing the multi-band reflectances in the training data, which has dimensions of $(n + 1) \times m$. The number of MODIS correlated bands is given by n. Because the uncertainties of MODIS bands 5–7 in the band conversion process are large, they are discarded in this approach. Only the first four MODIS bands are used in the current AB1 algorithm. The number of training data points in the grid is given by m, and **Y** is the matrix containing the albedo values in the training data and having a degree of $18 \times m$. Because the WSA and BSA from 0 to 80° in 5° increments must be calculated, there are 18 different albedo values. **A** is the matrix of regression coefficients, which has dimensions of $(n + 1) \times 18$. The uncertainties of MODIS bands 5–7 in the band conversion process are large and the inversion is sensitive to noise.

The least-squares solution A* can be expressed as follows:

$$\mathbf{A}^* = \left(\mathbf{X}^T\mathbf{X}\right)^{-1}\mathbf{X}^T\mathbf{Y} \tag{3.9}$$

The least-squares method normally has a simple form and provides good fitting results. However, if intercorrelations exist in the training dataset, a robust full retrieval cannot be made. Numerical tests indicate that if $\mathbf{A}^*$ is calculated by the simple least-squares solution, the result is sensitive to noise in the MODIS spectral data. Therefore, another technique was used here: a stable solution can be acquired by adding the simulated data noise into the regression algorithm. Specifically, $\mathbf{X}$ is the training data produced by the POLDER data after conversion to the MODIS bands. It is assumed that the training data are accurate but that the observations contain noise; therefore, statistical random noise is added to $\mathbf{X}$, which becomes $\tilde{\mathbf{X}}$. The anti-noise least-squares solution under this design can be expressed as follows, which is the least-error solution to be applied to observations with noise:

$$\mathbf{A}^* = \left(\tilde{\mathbf{X}}^T\tilde{\mathbf{X}}\right)^{-1}\tilde{\mathbf{X}}^T\mathbf{Y} \tag{3.10}$$

The procedure of adding noise to the data is not actually implemented when estimating $\mathbf{A}^*$ because the specific noise in limited data cannot represent the overall statistical properties of noise. Because $\tilde{\mathbf{X}}^T\tilde{\mathbf{X}}$ and $\tilde{\mathbf{X}}^T\mathbf{Y}$ can be directly calculated from X^TX and X^TY, it is possible to bypass generating $\tilde{\mathbf{X}}$. It is assumed that the noise components of the MODIS bands all average to zero and that their covariance matrix is Δ. Then,

$$\tilde{\mathbf{X}}^T\tilde{\mathbf{X}} = \mathbf{X}^T\mathbf{X} + m\Delta, \quad \tilde{\mathbf{X}}^T\mathbf{Y} = \mathbf{X}^T\mathbf{Y} \tag{3.11}$$

Therefore, the anti-noise least-squares solution can be calculated as

$$\mathbf{A}^* = \left(\mathbf{X}^T\mathbf{X} + m\Delta\right)^{-1}\mathbf{X}^T\mathbf{Y} \tag{3.12}$$

In the current algorithm, a simple estimate of the data noise in the MODIS bands has been incorporated, and Δ has been included as a diagonal matrix in which the diagonal elements are the variances of the noise in each band. Two main factors are embodied in Δ: the residual uncertainty of the atmospheric correction and the uncertainty introduced by the process of band conversion from POLDER to MODIS. At present, the noise components of the first two bands of MODIS are believed to be small enough that their standard deviation can be set to 0.01; the third and fourth bands are apt to be influenced by aerosol error, and therefore the standard deviation of their noises was set to 0.02. It is sometimes difficult to obtain a rigorous estimate of the constraints, such as data noise, in an inversion method. However, even if the constraints are not accurate, usually their negative effect on the inversion results will not be significant, but the benefit to the stability of the inversion will be apparent.

Table 3.3 The fitting residual of the algorithm applied to the training data

Category of training data	Average RMSE of WSA	Average RMSE of BSA at 45° solar zenith angle	Relative error of WSA (%)	Relative error of BSA at 45° solar zenith angle (%)
Vegetation	0.0078	0.0063	4.91	4.25
Bare ground	0.0118	0.0103	5.12	4.67
Ice/snow	0.0248	0.0199	3.85	3.19

To assess the regression results, the statistics of the inversion errors of WSA and BSA at the 45° solar zenith angle were calculated based on the training data. The average RMSE values of the 50061 ABs are listed in Table 3.3, which shows that the fitting performance of the training data with the regression equation is the best for a surface covered with vegetation. As the average albedo of bare ground and snowfield increases, the fitting residual also increases, while the relative error remains within 6 %; the WSA has the same variation trend as the **BSA.**

3.2.3 The AB2 Algorithm

The AB2 albedo inversion algorithm uses the TOA directional reflectance routinely acquired by the MODIS onboard Terra/Aqua platform to invert the surface broadband albedo directly. This algorithm does not require the atmospheric correction; thus, atmospheric correction difficulties and errors are avoided. The main difference between the AB1 and AB2 algorithms is that simulation of atmospheric radiative transfer is contained in the forward modeling procedure of the AB2 algorithm. Both algorithms have similar processing procedures except for the atmospheric simulation part. Because most steps are identical, they are not repeated in this section; instead, this section focuses on the atmospheric radiative transfer simulation.

The 6S atmospheric transfer code (Vermote et al. 1997, 2002) was used to simulate the TOA directional reflectances. To avoid extremely large computations, the 6S module was not used directly. Instead, a fast and accurate approximation method proposed by Qin et al. (2001) was chosen to simulate the TOA directional reflectances. This method is especially suitable for simulating atmospheric effects over non-Lambertian land surfaces. To eliminate the effects of gaseous absorption, the original equation was reformulated as:

$$
\begin{aligned}
&\rho^{TOA}(\theta_s, \theta_v, \varphi) \\
&= t_{H_2O}\left\{ \rho_0(\theta_s, \theta_v, \varphi) + \frac{T(\theta_s) \cdot R(\theta_s, \theta_v, \varphi) \cdot T(\theta_v) - t_{dd}(\theta_s) \cdot t_{dd}(\theta_v) \cdot |R(\theta_s, \theta_v, \varphi)| \cdot \overline{\rho}}{1 - r_{hh}\overline{\rho}} \right\}
\end{aligned}
$$

$$(3.13)$$

where the matrices $\mathbf{T}(\theta_s), \mathbf{R}(\theta_s, \theta_v, \varphi)$, and $\mathbf{T}(\theta_v)$ are defined as

$$\mathbf{T}(\theta_s) = [t_{dd}(\theta_s) \ t_{dh}(\theta_s)], \qquad \mathbf{T}(\theta_v) = \begin{bmatrix} t_{dd}(\theta_v) \\ t_{hd}(\theta_v) \end{bmatrix} \qquad (3.14)$$

$$\mathbf{R}(i, v) = \begin{bmatrix} r_{dd}(\theta_s, \theta_v, \varphi) & r_{dh}(\theta_s, \varphi_s) \\ r_{hd}(\theta_v, \varphi_v) & r_{hh} \end{bmatrix} \qquad (3.15)$$

where subscripts s, and v represent solar illumination and viewing direction, respectively; $\theta_s, \theta_v, \varphi_s, \varphi_v$, and φ are the solar zenith angle, view zenith angle, solar azimuth angle, view azimuth angle, and relative azimuth angle, respectively; $\rho^{TOA}(\theta_s, \theta_v, \varphi)$ is the TOA directional reflectance; $\rho_0(\theta_s, \theta_v, \varphi)$ is the path scattering reflectance of the atmosphere; t_{H_2O} is the transmittance of water vapor; $\bar{\rho}$ is the spherical albedo of the atmosphere; and t and r represent transmittance and reflectance, respectively. The subscripts h and d indicate hemispheric (diffuse) and directional (direct), respectively, and there are four combinations of these two symbols: dd, dh, hd, and hh with the first symbol indicating the initial status of photons (incoming) and the second indicating the resulting photon status after interaction (outgoing). Therefore, $t_{dd}(\theta_s)$ and $t_{dd}(\theta_v)$ are the downward and upward bidirectional path transmittances (BDTs), respectively; and $t_{dh}(\theta_s)$ and $t_{hd}(\theta_v)$ are the directional-to-hemispheric path transmittance (DHT) and the hemispheric-to-directional transmittance (HDT), respectively. If $\theta_s = \theta_v$, then $t_{dd}(\theta_s) = t_{dd}(\theta_v), t_{dh}(\theta_s) = t_{hd}(\theta_v), r_{dd}(\theta_s, \theta_v, \varphi), r_{dh}(\theta_s, \varphi_s), r_{hd}(\theta_v, \varphi_v)$, and r_{hh} are the BDR, directional-to-hemispheric reflectance (DHR), hemispheric-to-directional reflectance (HDR) and bi-hemispheric reflectance (BHR), respectively. If $\theta_s = \theta_v$ and $\varphi_s = \varphi_v$, then $r_{dh}(\theta_s, \varphi_s) = r_{hd}(\theta_v, \varphi_v)$. The values of $r_{dh}(\theta_s, \varphi_s), r_{hd}(\theta_v, \varphi_v)$, and r_{hh} were calculated using the POLDER-3/PARASOL BRDF database, and the values of $\rho_0(\theta_s, \theta_v, \varphi), t_{dh}(\theta_s), t_{hd}(\theta_v), t_{dd}(\theta_s), t_{dd}(\theta_v), \bar{\rho}$, and t_g were acquired using a look up table (LUT) produced by the 6S atmospheric transfer code.

The LUT has seven dimensions: atmospheric type, aerosol type, aerosol optical depth (AOD), target elevation, solar zenith angle, view zenith angle, and relative azimuth angle. The input parameters for the 6S code were set as follows (Table 3.4): six atmosphere types (tropical, mid-latitude summer, mid-latitude

Table 3.4 Input settings for parameters of the 6S atmospheric transfer code

Parameters	Input settings
Atmospheric type	Tropical, Mid-latitude summer, Mid-latitude winter, Subarctic summer, Subarctic winter, US62
Aerosol type	Continental, Maritime, Urban, Desert, Biomass burning, Haze
AOD	0.1, 0.2, 0.25, 0.3, 0.35, 0.4
Target altitude (km)	0, 0.5, 1.0, 1.5, 2.0, 2.5, 3, 3.5
Solar zenith (degree)	0, 4, 8, …, 76, 80
View zenith (degree)	0, 4, 8, …, 60, 64
Relative azimuth (degree)	0, 20, 40, …, 160, 180

winter, subarctic summer, subarctic winter, and US62 standard) and six aerosol types (continental, maritime, urban, desert, biomass burning, and haze (a user-defined aerosol type in which the percentages of four aerosol particles—dust, water soluble, soot, and oceanic—are 15, 75, 10, and 0 %)). The amount of water vapor was set to a default value, and the AOD at 550 nm was set to 0.1, 0.2, 0.25, 0.3, 0.35, and 0.4, which corresponds to a range from clear to relatively turbid aerosol-loading conditions. The target altitude varied from 0 to 3.5 km in increments of 0.5 km; the solar zenith angle varied from 0 to 80°, and the view zenith angle varied from 0 to 64°, both in 4° increments. The relative azimuth angle varied from 0 to 180° in increments of 20°.

3.2.4 The STF Algorithm

Intermediate products derived from the AB algorithms have greatly improved the temporal resolution of albedo data. However, intrinsic limitations still exist in these products. First, they suffer from frequent data gaps due to daily cloud cover and seasonal snow effect. Another limitation is the sharp fluctuations in albedo time series resulting from noises in reflectance inputs and the uncertainties of AB regression models. Therefore, a STF algorithm has been designed to address these issues.

As illustrated in Fig. 3.3, the procedure of the proposed STF algorithm consists of three components:

(1) calculation of the a priori statistics of global land surface albedo;
(2) statistics-to-model coefficients conversion;
(3) statistic-based temporal filtering of GLASS intermediate albedo product time series.

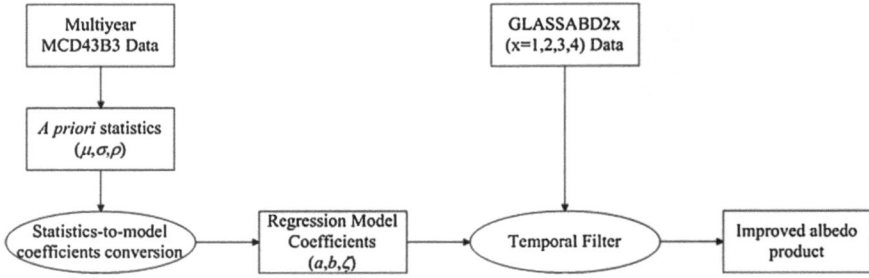

Fig. 3.3 Procedure of statistically based temporal filtering algorithm (Liu et al. 2013a)

3.2.4.1 Temporal Filtering Equation

Based on the temporal correlation of albedo measurements on consecutive days, it is reasonable to assume that the true albedo value a_k on the kth day is linearly correlated with the value $\alpha_{k+\Delta k}$ on the $(k + \Delta k)$th day:

$$\alpha_k = a_{\Delta k}\alpha_{k+\Delta k} + b_{\Delta k} + e_{\Delta k} \tag{3.16}$$

where both $a_{\Delta k}$ and $b_{\Delta k}$ are regression model coefficients. The model error $e_{\Delta k}$ is assumed to be Gaussian distributed with zero mean and variance $\zeta_{\Delta k}^2$ (see the next section for calculation of their values). Then, given the intermediate albedo product $\alpha_{k+\Delta k}^*$ and the corresponding uncertainty $\eta_{k+\Delta k}^2$ on the $(k + \Delta k)$th day, the probability density function (PDF) $P\left(\alpha_k|\alpha_{k+\Delta k}^*\right)$ of the kth day's true albedo α_k has the form $P\left(\alpha_k|\alpha_{k+\Delta k}^*\right) \sim N\left(a_{\Delta k}\alpha_{k+\Delta k}^* + b_{\Delta k}, \zeta_{\Delta k}^2 + a_{\Delta k}^2\eta_{k+\Delta k}^2\right)$. In other words, the predicted albedo on the kth day is $a_{\Delta k}\alpha_{k+\Delta k}^* + b_{\Delta k}$, and the corresponding prediction error consists of two components: the regression model error $\zeta_{\Delta k}^2$ and the propagated observational error $a_{\Delta k}^2\eta_{k+\Delta k}^2$. Specifically, when Δk equals zero, the PDF $P\left(\alpha_k|\alpha_k^*\right)$ of α_k is $N\left(\alpha_k^*, \eta_k^2\right)$. Moreover, the PDF $P\left(\alpha_k|\mu_k\right)$ of α_k given its a priori statistical mean μ_k and variance σ_k^2 (see the next section for calculation of these values) is $N\left(\mu_k, \sigma_k^2\right)$.

Assume that the PDFs $P\left(\alpha_k|\alpha_{k+\Delta k}^*\right)(\Delta k = -K, \ldots, K)$ are independent of each other. Then, given the surrounding intermediate products $\alpha_{k+\Delta k}^*(\Delta k = -K, \ldots, K)$ and the statistical mean μ_k, the joint PDF $P\left(\alpha_k|\alpha_{k-\Delta K}^*, \ldots, \alpha_{k+\Delta K}^*, \mu_k\right)$ can be expressed as:

$$P\left(\alpha_k|\alpha_{k-\Delta K}^*, \ldots, \alpha_{k+\Delta K}^*, \mu_k\right) = P\left(\alpha_k|\mu_k\right)\prod_{\Delta k=-K}^{\Delta k=+K} P\left(\alpha_k|\alpha_{k+\Delta k}^*\right) \tag{3.17}$$

Consequently, the maximum likelihood estimate of the kth day's true albedo α_k takes the form:

$$\widehat{\alpha_k} = \left(\frac{\mu_k}{\sigma_k^2} + \sum_{\Delta k=-K}^{\Delta k=+K} \frac{a_{\Delta k}\alpha_{k+\Delta k}^* + b_{k+\Delta k}}{\zeta_{\Delta k}^2 + a_{\Delta k}^2\eta_{k+\Delta k}^2}\right)c \tag{3.18}$$

where c is the corresponding prediction variance:

$$c = 1 \left/ \left(\frac{1}{\sigma_k^2} + \sum_{\Delta k=-K}^{\Delta k=+K} \frac{1}{\zeta_{\Delta k}^2 + a_{\Delta k}^2\eta_{k+\Delta k}^2}\right)\right. \tag{3.19}$$

Equation (3.18) shows that the STF method is essentially a weighted average of statistical and observed values. The STF method has some advantages over previous approaches (Fang et al. 2007; Moody et al. 2005). First, the weights in the STF equation depend on both the temporal correlation and the observational errors of albedos for consecutive days. Observations with better correlation and less noise contribute more to the filtered result. Second, the STF method can improve both the smoothness and the integrity of albedo time series. The observational albedo on the kth day is smoothed if it is valid; if not, the missing value is filled in. Third, the statistical mean is used to fill in data gaps if all surrounding observations are absent. Finally, the STF method provides an assessment of the uncertainty of filtered results.

3.2.4.2 The a Priori Knowledge Dataset of Global Surface Albedo

A *priori* knowledge is of critical importance in retrieving BRDF/albedo parameters (Li et al. 2001; Liang 2008). In the present study, the prior statistics of global surface albedo were calculated from multiyear MCD43B3 products (2000–2009). There are two reasons to use MCD43B3 products: (1) they use the same inputs as the AB algorithm and (2) they provide excellent temporal stability.

The a priori statistics derived from MCD43B3 products include: (1) the multiyear albedo mean μ_k and standard deviation $\sigma_k (k = 1, 9, \ldots, 361)$ and (2) the correlation coefficient (CC) $\rho_{\Delta k}(\Delta K = -32, -24, \ldots, 24, 32)$ for two consecutive days' albedos. Some regions may be missing valid retrievals over multiple years (e.g., tropical regions persistently covered by cloud or polar regions affected by polar night). For polar areas, the average of two valid albedo means before polar night is used to fill albedo mean gaps. For other regions, albedo means for each class at each 10° latitude band are calculated and then used to fill albedo mean gaps according to MODIS IGBP classification products (MCD12Q1). In consideration of the storage consumption and the stability of the a priori statistics, the a priori knowledge dataset has a 5 km spatial resolution and an 8-day temporal resolution.

To obtain the daily statistics and model coefficients required by Eq. (3.18), the 8-day a priori statistics must be interpolated to a daily resolution. To this end, a cubic polynomial function is applied to the statistical mean and standard deviation; and an exponential function is used for CCs. The curve-fitting functions are as follows:

$$
\begin{cases}
\mu(\Delta d) = \lambda_1 \Delta d^3 + \lambda_2 \Delta d^2 + \lambda_3 \Delta d + \lambda_4 \\
\sigma(\Delta d) = \lambda_5 \Delta d^3 + \lambda_6 \Delta d^2 + \lambda_7 \Delta d + \lambda_8 \\
\rho(\Delta d) = \exp\left(\lambda_9 \Delta d^4 + \lambda_{10} \Delta d^2\right)
\end{cases}
\tag{3.20}
$$

Once the fitting coefficients $\Lambda = \{\lambda_1, \lambda_2, \ldots, \lambda_{10}\}$ have been obtained, the daily statistical mean, standard deviation and CCs can be calculated from Eq. (3.20). Then a statistics-to-model coefficients conversion procedure is used to calculate the model coefficients in Eq. (3.18):

$$\begin{cases} a_{\Delta k} = \rho_{\Delta k} \dfrac{\sigma_k}{\sigma_{k+\Delta k}} \\ b_{\Delta k} = \mu_k - a_{\Delta k}\mu_{k+\Delta k} \\ \zeta_{\Delta k}^2 = \left(1 - \rho_{\Delta k}^2\right)\sigma_k^2 \end{cases} \tag{3.21}$$

3.3 Product Characteristics, Quality Control and Validation

3.3.1 Product Characteristics

There are two categories of GLASS albedo products: intermediate products and final products. The final products have the best quality and could be provided to public data users. Although the intermediate products are output from the production chain, they are also plagued with serious flaws and overabundant information. A complete list of all of the intermediate and final albedo products in GLASS is presented in Table 3.5. However, only the final products are recommended to public users. As user-oriented remote sensing products, the GLASS final albedo products have distinguishing features of spatial and temporal completeness, long-term consistency, and accuracy.

3.3.2 Quality Control and Assessment

Every scene of the GLASS final albedo products has been subjected to a strict QC procedure. QC has been carried out both automatically and with human involvement. Automatic QC refers mainly to the generation of the QC flag which estimate the uncertainty of the GLASS final albedo products. The quality check with human involvement is a computer-aided visual inspection of the spatial and temporal patterns, of the product before its final release.

3.3.2.1 Quality Control Flag

GLASS albedo products are composed of shortwave white-sky albedo, shortwave black-sky albedo at local noon solar angle, and the QC flag. The QC flag gives a pixel-wise description of the data processing parameters as well as the credibility of the result. The flag is a 16-bit data field provided for each pixel. The bitwise interpretation of the QC flag for the GLASS02A05 and GLASS02A06 products is given in Table 3.6, and the details of this interpretation can be found in the user manual. The lowest two bits give an indication of the overall quality of the albedo

Table 3.5 List of GLASS intermediate and final albedo products

Product	Type	Algorithm	Input	Temporal resolution (day)	Composition interval (day)	Projection	Spatial resolution
GLASS02A21	I	AB1	MOD09GA	1	1	SIN	1 km
GLASS02A01	I	Average	GLASS02A21	8	17	SIN	1 km
GLASS02A22	I	AB1	MYD09GA	1	1	SIN	1 km
GLASS02A02	I	Average	GLASS02A21	8	17	SIN	1 km
GLASS02A23	I	AB2	MOD021 km	1	1	SIN	1 km
GLASS02A03	I	Average	GLASS02A21	8	17	SIN	1 km
GLASS02A24	I	AB2	MYD021 km	1	1	SIN	1 km
GLASS02A04	I	Average	GLASS02A21	8	17	SIN	1 km
GLASS02A06	Final	STF	GLASS02A21 GLASS02A22 GLASS02A23 GLASS02A24	1	17	SIN	1 km
GLASS02B06	Final	Upscaling	GLASS02A06	8	17	CMG	0.05°
GLASS02B05	Final	AB1 + STF	LTDR AVHRR dataset	8	33	CMG	0.05°

I: intermediate products

Table 3.6 Quality Control (QC) flags of the final GLASS albedo product

Bit No.	Parameter Name	Value/State	
0–1	Overall quality	00: good	01: acceptable
		10: with uncertainty	11: prior value
2–3	Land cover state	00: vegetation	01: bare ground
		10: snow	11: un-classified
4–5	Length of composite window	00: 8-day	01: 16-day
		10: 24-day	11: 32-day
6–8	Number of actually used (clear-sky) intermediate products	000: 0	001: 1
		010: 2–3	011: 4–7
		100: 8–15	101: 16–31
		110: 32–63	111: 64–127
9–10	Ratio of actually used to total number of intermediate products	00: more than 50 %	01: 25–50 %
		10: 10–25 %	11: less than 10 %
11–14	Uncertainty of albedo retrieval	0000: 0.00–0.01	0001: 0.01–0.02
		0010: 0.02–0.03	0011: 0.03–0.04
		0100: 0.04–0.05	0101: 0.05–0.06
		0110: 0.06–0.07	0111: 0.07–0.08
		1000: 0.08–0.09	1001: 0.09–0.10
		1010: 0.10–0.11	1011: 0.11–0.12
		1100: 0.12–0.13	1101: 0.13–0.14
		1110: 0.14–0.15	1111: >0.15
15	Albedo validity flag	0: valid value	1: invalid value

product in the pixel, with '00' indicating a "good" estimation result with an uncertainty of less than 0.01 absolutely or 5 % relatively, '01' indicating an "acceptable" estimation with an uncertainty of less than 0.05 absolutely or 10 % relatively, '11' indicating the most uncertain estimation in which the a priori value has been used to fill the pixel, and '10' indicating a state between "acceptable" and a "fill value". In addition to the overall quality assessment, the uncertainty of the albedo estimate is also quantitatively given in bits 11–14. This uncertainty estimate is first generated in the AB1 and AB2 algorithms and passed to the STF algorithm to derive the final uncertainty using statistical principles. However, this uncertainty is only a statistical estimate and does not always reflect the actual error in a specific case.

To understand the spatial and temporal distribution of the quality of the GLASS albedo product, the seasonal variation in the quality flag at typical sites was first investigated. Histograms of the overall quality flag were derived in monthly temporal windows and 7×7-pixel spatial windows around six typical FLUXNET sites, the names and basic information of which are listed in Table 3.7. The statistics are presented as a percentage bar in Fig. 3.4. The yellow bar indicates the percentage of pixels with the overall quality flag '11', which represents the lowest quality. It can be observed that these yellow bars occur at the RU-Che, DE-Hai, and BR-Cax sites. In winter and in high-latitude areas, it is either polar night or the solar zenith angle is greater than 80°; therefore, there is no valid observation in the

Table 3.7 Information about six typical FLUXNET sites on different continents

Site_name	Latitude	Longitude	Cover_type
AU-Tum	−35.6557	148.152	Evergreen broadleaf forest
BR-Cax	−1.71972	−51.459	Evergreen broadleaf forest
DE-Hai	51.0793	10.452	Deciduous broadleaf forest
RU-Che	68.6147	161.339	Mixed forest
US-Fpe	48.3077	−105.1019	Grassland
ZA-Kru	−25.0197	31.4969	Savanna

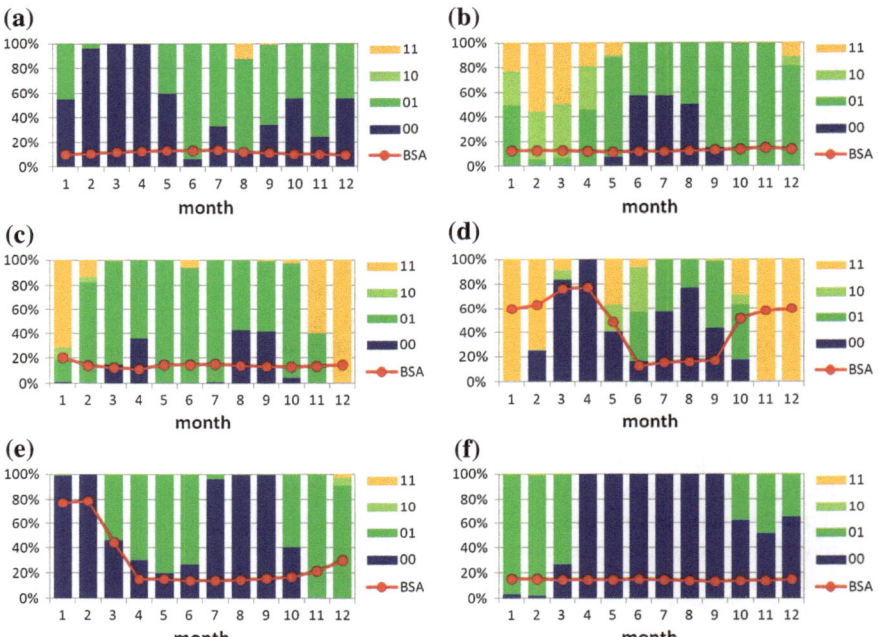

Fig. 3.4 Monthly average percentages of GLASS albedo products with various overall quality flags in 2004. **a** AU-Tum; **b** BR-Cax; **c** DE-Hai; **d** RU-Che; **e** US-Fpe; **f** ZA-Kru. (Liu et al. 2013b)

MODIS shortwave bands to provide information about surface albedo. This fact explains the low quality of the products for the RU-Che and DE-Hai sites. In the case of the BR-Cax site, there are always heavy clouds from November to May in the tropical forest area; therefore, many of the pixels could not have a cloud-free observation within the 16-day composition window and were filled with an a priori value. The blue, dark-green, and light-green bars indicate the data with "good," "acceptable," and "with uncertainty" quality flags, respectively. Most of the "good" data can be found in mid-latitude areas such as the US-Fpe, AU_Tum, and ZA_Kru sites. The pixels marked as "acceptable" and "with uncertainty" also indicate cloud contamination, which mostly occurs in the rain/snow seasons.

To obtain a clearer view of the spatial distribution of the quality of the GLASS albedo product, maps of the yearly average percentage of good-quality data and the average uncertainty in the global scope have been derived. Figure 3.5a presents the spatial distribution of the average percentage of "good" quality pixels in the final GLASS albedo product for 2003 and 2004. A large part of the land areas at low latitudes have approximately 100 % "good" quality product, with exceptions in the tropical rain forests of Amazonia, central Africa, and Indonesia and the monsoon areas in India and Southeast Asia, where cloud cover is frequent throughout the rainy season or the whole year. In high-latitude areas, the percentage of good-quality pixels decreases with latitude. This decrease is due to the phenomena of large solar zenith angle and polar night. Horizontal strips can be seen in high-latitude areas because if the average solar zenith angle of a MODIS tile is greater than a threshold (80°), the entire tile is excluded from the calculation. Figure 3.5b presents the spatial distribution of the average uncertainty in the final GLASS albedo product for 2003 and 2004. The pattern is essentially the inverse of Fig. 3.3a because there are fewer good-quality pixels where large uncertainties occur. It is apparent that, except for the polar areas, the yearly average uncertainty is less than 0.05 in most areas.

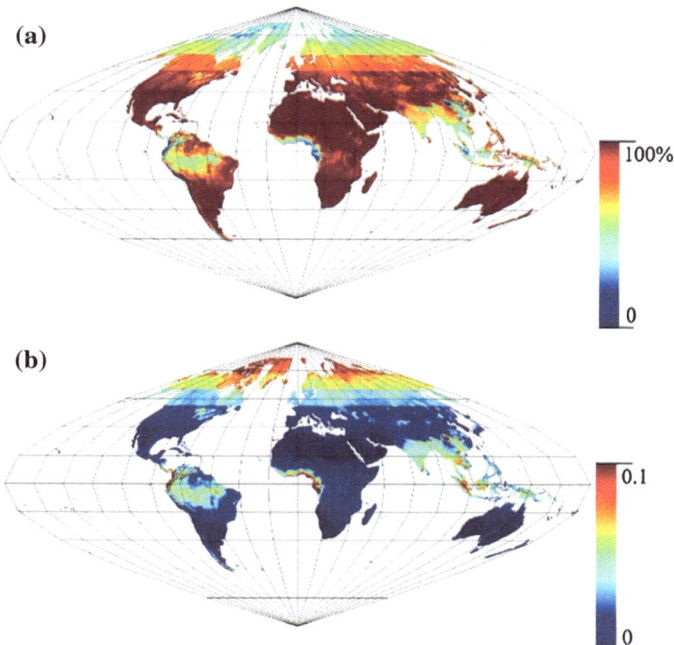

Fig. 3.5 Spatial distribution of the average quality of the GLASS albedo product in 2003 and 2004. **a** percent of the product with a "good" quality flag; **b** average product uncertainty (Liu et al. 2013b)

To view the interannual variation in the quality flag of the GLASS albedo
product, 402 sample sites scattered all around the world were selected and the
statistics of the quality flags for these sites were derived. These sites form part of
the CEOS-Benchmark Land Multisite Analysis and Intercomparison of Products
(BELMANIP) network (Baret et al. 2006). Figure 3.6 shows the locations of these
sites overlaid on top of a global land/sea boundary map. Figure 3.7 shows the
seasonal average percentage of the quality flags for all sites from 1982 to 2010.
The percentages of good-quality products are significantly larger in 2000–2010
than in 1982–1999 because the former is a MODIS-derived product and the latter
is AVHRR-derived. The quality improves more after 2003 because the successful
launch of the Aqua satellite doubled the number of the available MODIS observations. In the latter half of 1994 (October 1994 to January 1995), there is a gap
where no GLASS albedo product was produced because the LTDR version-3
archive does not provide AVHRR data for this period between NOAA 11 and
NOAA 14.

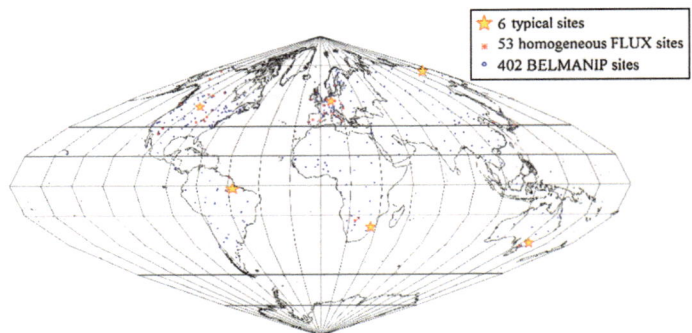

Fig. 3.6 Locations of the sample sites used (Liu et al. 2013b)

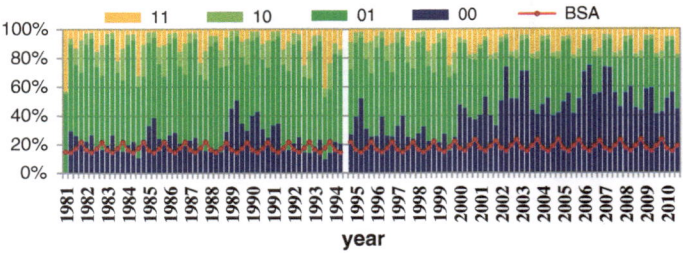

Fig. 3.7 Interannual variation of the quality flags for the GLASS albedo product, averaged over
all BELMANIP sites every three months (Liu et al. 2013b)

3.3.2.2 Visual Inspection

The quality control process with human involvement required individuals with special expertise to review the statistics, maps, and diagrams visually for every scene of GLASS albedo products. The purpose of visual inspection was to ensure that users are provided with high-quality products. Visual inspection consists of two parts: a spatial quality check and a temporal dynamics check.

3.3.2.3 Spatial and Temporal Completeness

During the quality control process, every scene of GLASS albedo products was visualized in a computer interface and checked for missing data or bands, invalid values, spatial discontinuities, and any abnormal patterns arising from the lack of raw data or any other problems. If any undesirable flaws are found, a feedback process is triggered, and the problematic products are regenerated until the flaws have been appropriately dealt with.

This inspection proved that the GLASS albedo products have satisfactory spatial and temporal completeness and continuity. As two examples, Fig. 3.8 gives the global map of GLASS02B06 black-sky albedo for day 209, 2008, and Fig. 3.9 gives maps of the GLASS02A06 black-sky albedo of tile H25V05 in the MODIS sinusoidal grid, for days 209–361, 2008 (every 8 days). This area is located on the Tibetan Plateau and is covered by grassland, bare soil, lakes, and snow. It can be seen from Figs. 3.8 and 3.9 that there are no spatial or temporal gaps in the GLASS02A06/GLASS02B06 albedo maps. The spatial distribution of the different

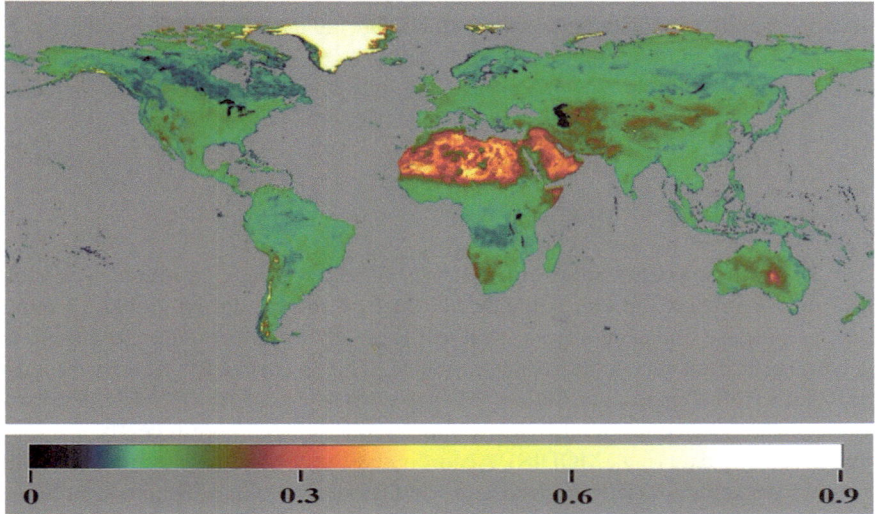

Fig. 3.8 Global view of GLASS02B06 black-sky albedo on day 209, 2008

GLASS02A06

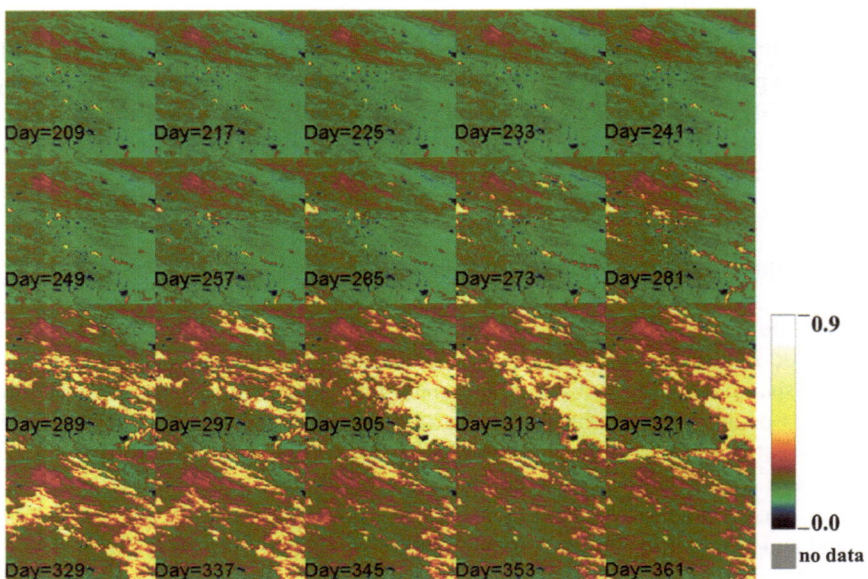

Fig. 3.9 Consecutive maps of GLASS02A06 black-sky albedo of tile H25V05 (Liu et al. 2013a)

land covers is well reflected: bare soil areas are in red, grasslands areas are in green, inland waterbodies are in dark blue, and snow-covered areas are in yellow. Seasonal variations in albedo are well depicted by the consecutive GLASS02A06 albedo maps shown in Fig. 3.2. During days 209–241 (July 27–August 28), the albedo variability in this area was very small. During days 249–305 (September 5–October 31), the snow areas increased gradually as winter approached. During days 313–361 (November 8–December 26) snow areas decreased with increased snow melt.

3.3.2.4 Long-Term Consistency

The 30-year long-term series of GLASS albedo products provides valuable information to study the energy balance of the land surface during global warming. For this purpose, the accuracy, as well as the ability to reflect temporal variations of land surface albedo is of great importance. This is especially the case because the GLASS albedo product before 2000 is derived from the AVHRR historical dataset, which has considerably lower quality than other, more recently acquired remote sensing data (e.g., MODIS). Accordingly, the consistency of the albedo and its trend between AVHRR-derived and MODIS-derived GLASS albedo products must be checked. In the quality control process, the long-term albedo series of typical sites, as well as some regional and global statistics, have been visualized as curve diagrams, inspected by experts, and cross referenced to other historical records.

It was found that the GLASS albedo products have maintained a good balance between temporal resolution and temporal continuity: seasonal variations of albedo are precisely reflected while false fluctuations due to imperfect atmospheric correction are smoothed out. Moreover, except for a decreased temporal resolution in the AVHRR-derived product, the albedo series before and after 2000 are consistent with each other. Figure 3.10 shows some examples of long-term albedo values from both AVHRR and MODIS data at some typical sites. The results indicate that the retrieved long-term albedo record is very stable.

3.3.3 Validation

The FLUXNET network (Baldocchi et al. 2001) is currently the largest global dataset of energy and mass flux measurements at the ecosystem scale. In a recent approach (Cescatti et al. 2012), the "La Thuile" FLUXNET database (www.fluxdata.org, October 2010) has been compared with MCD43A1, the MODIS BRDF/albedo product, which has a 500 m spatial resolution and an 8-day temporal step, and the consistency between these datasets for different areas and plant functional types has been analyzed. In this approach, 53 FLUXNET sites were considered to meet the criterion of land cover homogeneity. To minimize the uncertainty caused by the discrepancy in the observation scale, the same 53 homogeneous sites in the "La Thuile" database were also used to validate the 1 km-resolution GLASS albedo products. When processing the ground measurements, the actual albedo, which is the ratio of the upward to the downward shortwave radiant flux measured by tower-based pyranometers within two hours around local noon, was extracted from the database. Invalid values or measurements with insufficient incoming radiant flux, which are mostly caused by clouds and can be identified when the downward radiant flux is less than 70 % of the clear-sky irradiance, were screened out. The clear-sky radiance was calculated using a 6S atmospheric radiative transfer model (Vermote et al. 1997, 2002) with the default atmosphere parameters of a mid-latitude summer and a solar angle corresponding to the measurement time and location.

The 53 homogeneous FLUXNET sites selected by the approach described above (Cescatti et al. 2012) were used to validate the accuracy of the GLASS albedo product. The albedo measurement from the flux tower is affected by atmospheric conditions and has less credibility when the incoming solar irradiance is blocked by clouds. Therefore, the observations with a downward irradiance of less than 70 % of the clear-sky irradiance, which are approximately 45 % of the total valid data for these sites were screened out. The blue-sky albedo of the GLASS product was then calculated as a combination of BSA and WSA weighted by a sky light ratio factor, which was simply estimated as the function of the solar angle given in (Long and Gaustad 2004). Because only clear-sky ground

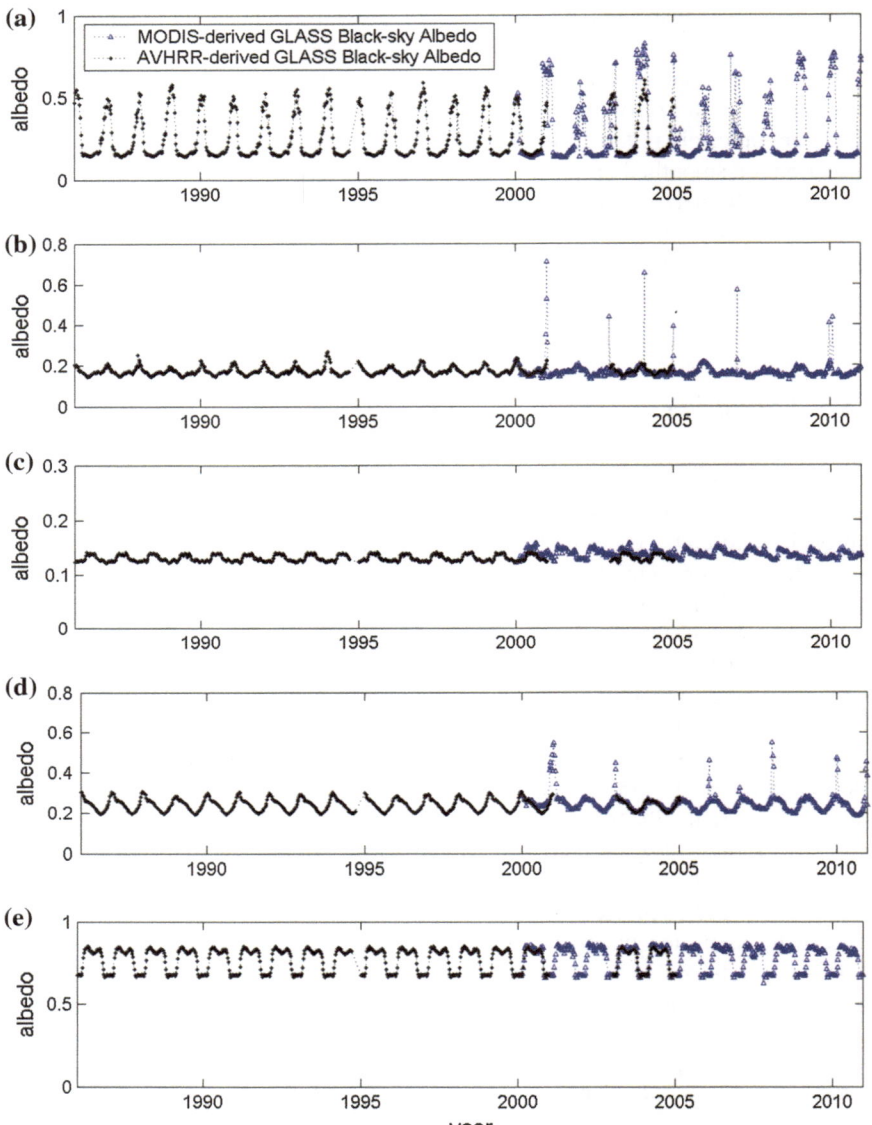

Fig. 3.10 Long-term GLASS albedo product at five sites: **a** US-Fpe, 48.3079 N, −105.101E, grassland; **b** ARM_SGP_Main, 36.605 N, −97.4884E, cropland; **c** Duke_forest_hardwoods, 35.9736E, −79.1004 N, mixed forest; **d** Naiman_site, 42.9333E, 120.700 N, desert; **e** NASA-SE, 66.4797E, −42.5002 N, ice sheet. From Liang et al. (2013b), Int. J. Digit. Earth. Copyright © 2013 reprinted by permission of Taylor and Francis Ltd

measurements were used to validate the GLASS product, the error resulting from this simplistic estimation of the sky light ratio was acceptable.

Figure 3.11 presents the time series of the ground measurements and the GLASS blue-sky albedo for the six typical sites. The black dots indicate the clear-sky ground measurements, whereas the green dots indicate the ground

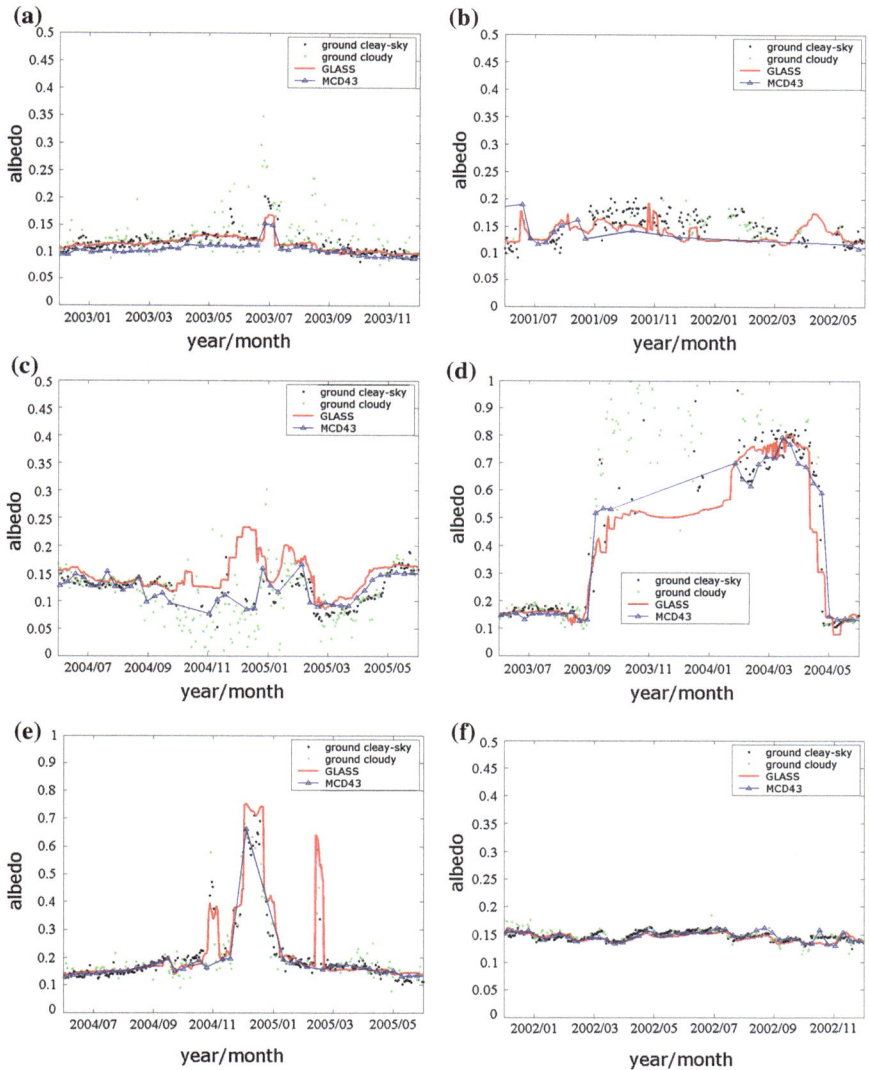

Fig. 3.11 Comparison of a one year time series of the ground-measured albedo and the GLASS blue-sky albedo for the typical FLUXNET sites. **a** AU-Tum; **b** BR-Cax; **c** DE-Hai; **d** RU-Che; **e** US-Fpe; **f** ZA-Kru (Liu et al. 2013b)

measurements that were screened out. The thick red solid lines indicate the blue-sky albedo extracted from the GLASS02A06 daily product. The blue solid lines with triangular markers indicate the blue-sky albedo from the MCD43BA 8-day product. Generally, the GLASS product reflects the trend and magnitude of the ground observations. The accuracy is better for clear days than for cloudy days. However, some general underestimation of the albedo can be found in winter when snow is most likely present and clear-sky days are rare. Almost no bias was found in snow-free seasons, except at the DE-Wet site where the values of the GLASS product were slightly higher, which may have been caused by the scale discrepancy. The footprint size of the surface albedo measurements (mostly 10–20 m) is considerably different from that of the MCD43A3 product (500 m) or the GLASS02A06 product (1 km). These 53 selected FLUXNET sites were relatively homogeneous in a region up to 2 km around each site. The scale discrepancy was minimized, but it still exists. Smoothing effects can also be observed in the GLASS product; although the GLASS algorithm tries to preserve the details of the temporal variation of the true surface albedo, there is still a compromise between the actual temporal resolution and noise sensitivity.

A scatter plot of all the clear-sky observations for these 53 homogeneous sites is given in Fig. 3.12a, and the statistics are given in Table 3.8. To analyze the error statistics of the GLASS product further, the GLASS quality flags were investigated using the observations only with the "good" quality flags and ground measurements for the corresponding date in the clear-sky dataset. The scatter plot for these "good" observations is given in Fig. 3.12b. The points are clearly closer to the diagonal line, and the Root Mean Square Deviation (RMSD) drops from 0.0587 to 0.0455, which means that the quality flag is a pertinent indicator of the accuracy of the GLASS product.

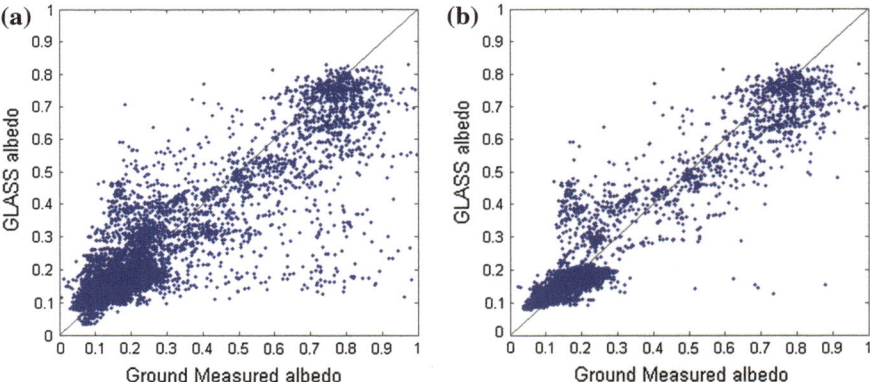

Fig. 3.12 Scatterplot of the FLUXNET ground measurements and the blue-sky albedo extracted from the GLASS product **a** clear-sky ground measurements and all GLASS data; **b** clear-sky ground measurements and "good" quality GLASS data (Liu et al. 2013b)

Table 3.8 Statistics comparing the FLUXNET ground measurements with the GLASS albedo product

Sample criterion	Number of Observations	Bias	RMSD	R^2
Clear-sky ground measurements and all GLASS data	28881	−2e-06	0.058	0.803
Clear-sky ground measurements and "good" quality GLASS data	16960	−0.001	0.045	0.893
Vegetation	14833	−0.001	0.030	0.617
Bare ground	938	0.012	0.053	0.309
Snow/ice	1189	0.004	0.125	0.723

The GLASS albedo algorithm follows a very simple classification rule when generating the product. The land surface state is divided into three categories by the band threshold: vegetation (when NDVI is greater than 0.2), snow/ice (when the reflectance in the blue band is greater than 0.3), and bare ground (other cases). The statistics of these three categories of GLASS albedo are given in Table 3.8. Clearly, the RMSD is the smallest for the vegetation observations and the largest for the snow/ice-covered observations. Also note that most of the FLUXNET observations are in the vegetated state. The validation for bare ground or snow/ice-covered surfaces is not sufficient, and the RMSD is larger.

The preliminary validation and evaluation work was still limited by data sources as well as methodology. There are never good-quality ground measurements when a global dataset needs to be validated. Measurements for homogeneous sites that are representative to a pixel size of 1 km are even scarcer. Although many researchers suggest using an upscaling method to validate remote sensing products for heterogeneous pixels, these procedures are not yet operational. Therefore, the results given in the direct validation section here cannot be looked upon as strict estimate of the absolute accuracy of the GLASS product, but only as an indication.

3.4 Preliminary Analysis

3.4.1 Global and Zonal Statistics and Trend Analysis

As a demonstration of possible applications, a time series of the global average of black-sky albedo was extracted from the GLASS product. The average land surface albedo values and their anomalies are shown in Fig. 3.13. The average albedo value over land surfaces is approximately 0.2 excluding the Antarctic and Arctic regions. There are significant seasonal variations. Although there is no overall trend, the decreasing trend after 2005 and the increasing trend from the end of the 1980s and the early 1990s can be clearly identified.

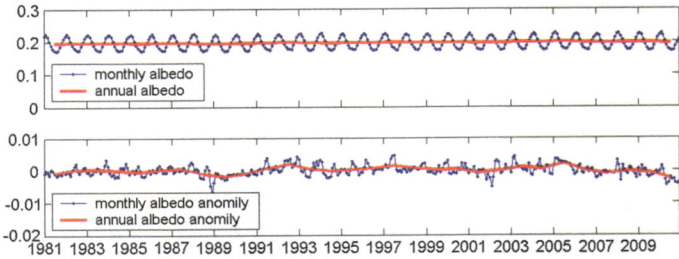

Fig. 3.13 Global land surface average albedo values and anomalies

Fig. 3.14 Spatial and temporal variations in global land surface albedo anomalies

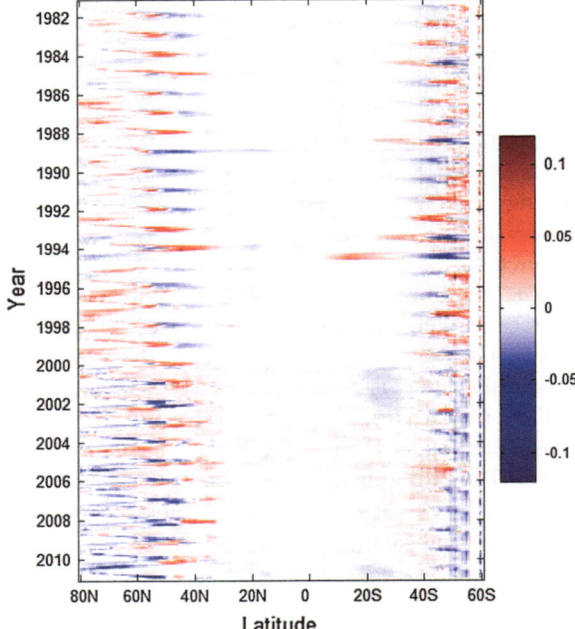

The detailed spatial and temporal variations of global albedo from the GLASS albedo product are shown in Fig. 3.14. Multiple factors affect these variations at different spatial and temporal scales.

The regional averages of albedo were calculated in four latitudinal zones: S60-S30, S30-N30, N30-N60, and N60-N90. Figure 3.15 presents the average albedo in these latitudinal zones. To provide a clear illustration, the albedo time series were temporally aggregated into monthly and yearly averages. The yearly average albedo in each latitudinal zone did not change significantly from 1982 to 2010, but the seasonal pattern shows some interannual change, especially in the third (N30-N60) zone. In the first (S60-S30) and second (S30-N30) zones, the seasonal

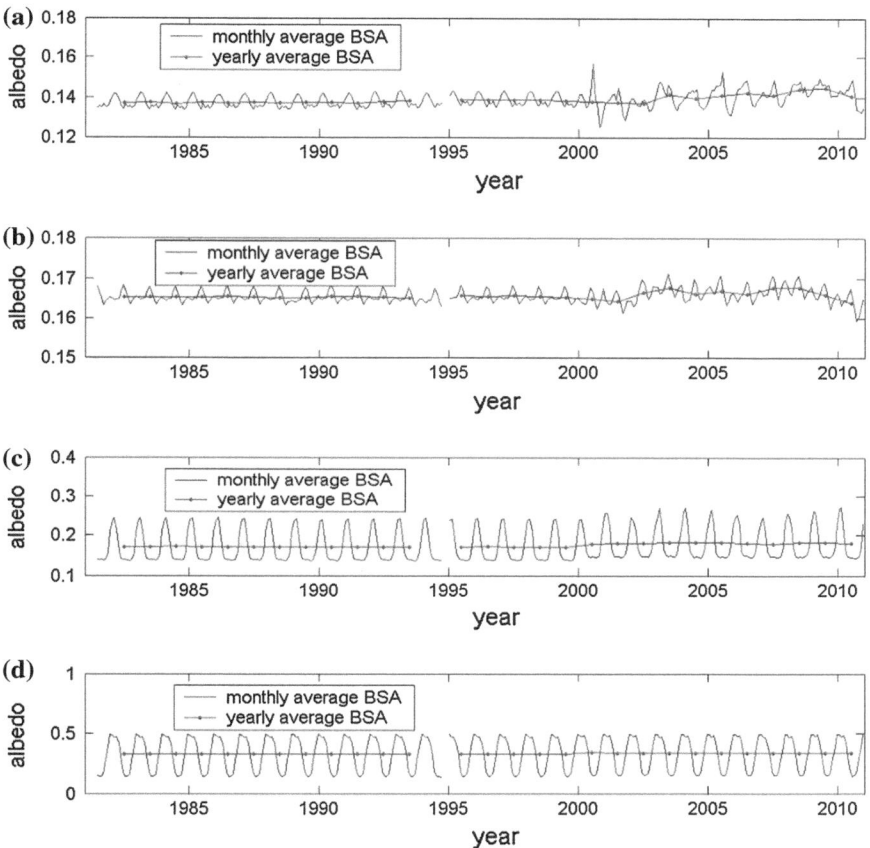

Fig. 3.15 Time series of the average black-sky albedo of different latitudinal zones, extracted from the GLASS product **a** S60-S30; **b** S30-N30; **c** N30-N60; **d** N60-N90 (Liu et al. 2013b)

variation seems to be intensified after 2000, which is most likely an artificial phenomenon due to the difference between the AVHRR-derived albedo and the MODIS-derived albedo. As shown in Fig. 3.5, the low-quality GLASS albedo values are distributed in the tropical area, i.e., the second zone, due to cloud contamination. The albedo products in cloud-contaminated areas obtain very little information from the AVHRR observations, but comparatively more information from the MODIS observations. This discrepancy explains why the variation of the albedo product intensified after 2000. Furthermore, the dynamic ranges for the vertical axes of Fig. 3.15a, b are very small, which also amplifies the apparent difference. The largest standard deviation of the yearly average albedo occured in the third zone (N30-N60). However, unlike the large seasonal variation, the interannual variation is not easily detectable by eye.

Table 3.9 Results of the Mann–Kendall trend test of the yearly average albedo values in different latitudinal zones

Year range	Latitude Zone	S60-S30	S30-N30	N30-N60	N60-N90
1982–2010	Z_s	4.7811	2.2918	3.4376	3.5167
	Standard deviation	0.0022679	0.00091598	0.0051761	0.0034161
	Trend/significance	I/Y	I/Y	I/Y	I/Y
	Trend anomaly year	1994	1997	2001	1998
1982–1999	Z_s	3.5426	1.2358	−0.8239	0.4943
	Standard deviation	0.00066697	0.00018284	0.00038373	0.00043273
	Trend/significance	I/Y	I/N	D/N	I/N
	Trend anomaly year	1992	–	–	–
2000–2010	Z_s	2.2576	0.0778	0.5449	−1.6348
	Standard deviation	0.0026565	0.0013428	0.0015823	0.0022592
	Trend/significance	I/Y	I/N	I/N	D/N
	Trend anomaly year	2005	–	–	2001

According to the principle of the Mann–Kendall test, the increasing trend is significant at greater at 95 % confidence level when Z_s is greater than 1.96, and the decreasing trend is significant at 95 % confidence level when Z_s is less than −1.96. Here, I means increasing, D means decreasing; Y means significant, and 'N' means insignificant

To analyze the changing trends in yearly average albedo, the Mann–Kendall trend test (Kendall 1976; Mann 1945) was used to determine the significance of the trend and to find trend anomalies. The results of this analysis are presented in Table 3.9. When the years from 1982 to 2010 were considered, a significant increasing trend was detected in all zones, and trend anomalies were detected near 2000 in three zones. This result indicates that the switch of the source data from AVHRR to MODIS had an influence on the trend analysis. As a compromise, the time range was split into two parts, 1982–1999 and 2000–2010, and the trend in each part was analyzed separately. These results also appear in Table 3.9. In the split time ranges, there are no significant trends except for the increasing trend in the first zone (S60-S30), and possible (not significant) decreasing trends in the third zone (N30-N60) during 1982–1999 and in the fourth zone (N60-N90) during 2000–2010. The increase in albedo in the Southern Hemisphere and the decrease in albedo in the Northern Hemisphere are in accordance with other literature (Zhang et al. 2010).

3.4.2 Regional Analysis

A warming trend in the Arctic since the early 1980s has been identified from satellite observations (Comiso 2003), and massive melt events over Greenland have been observed in satellite data for the last three decades (Abdalati and Steffen 1997; Mote 2007; Nghiem et al. 2012). Snow and ice surfaces normally have high reflectivity, but the melting events can reduce this reflectivity significantly. Through a positive feedback mechanism, melting of the Greenland ice sheet plays a key role in climate change and serves as an early sign of global warming. In addition, research has shown that high-latitude summer warming can offset the cooling effects caused by increased cloud coverage (Chapin et al. 2005). The long-term consistent GLASS surface albedo products provides a chance to revisit the surface changes over Greenland since the early 1980s. He et al. (2012) investigated the 32-year summer albedo change both temporally and spatially on the basis of the GLASS albedo products (the albedo products for the latest 2 years over Greenland were produced specifically for this study).

From 1981 to 2012, surface albedo for Greenland decreased at a rate of 0.0004 yr^{-1} ($p < 0.1$) (Fig. 3.16). Before 2000, the albedo changed annually, reaching maximum and minimum values in 1992 and 1994, respectively. However, the general trend before 2000 is not quite significant. Surface albedo has decreased 0.0024 yr^{-1} ($p < 0.05$) since 2000, which is mainly attributable to multiple melting events in Greenland. Previous researches have linked climate change in Greenland with regional/global climate indices in an effort to find the reason for the warm summer in recent years (Box et al. 2012; Stroeve 2001). A correlation ($r = 0.45$) was found between the North Atlantic Oscillation (NAO) index and the 32-year Greenland surface albedo anomaly (shown in Fig. 3.16). This result might be explained by noting that the negative summer NAO index

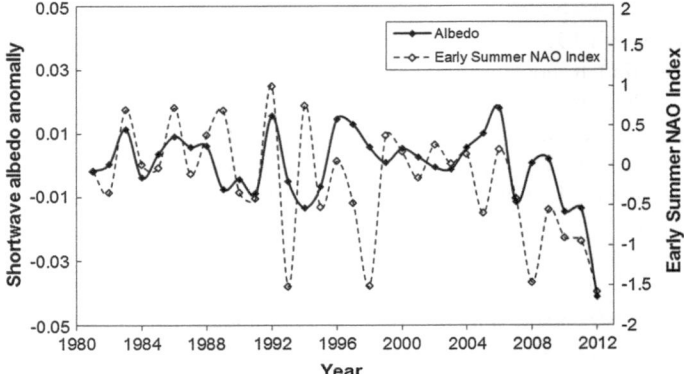

Fig. 3.16 Surface albedo changes over whole of Greenland from GLASS data and early summer (average of may, june, and july) NAO index (1981–2012)

values are related to melting and the positive NAO values to precipitation in Greenland (Box et al. 2012; Bromwich et al. 1999).

Various trends in albedo change were found also in the spatial domain. Greenland was divided into seven regions on the basis of Global Multi-resolution Terrain Elevation Data 2010 (GMTED2010) (Danielson and Gesch 2011) (Fig. 3.15). Statistical analysis showed significantly different trends before and after 2000 in each elevation range (Table 3.10). Before 2000, areas below 2,000 m did not show a statistically significant trend in albedo change, while those over 2,000 m showed a slight increase in albedo because of snow accumulation. Surface albedo decreased in almost all elevation ranges after 2000, with the most significant changes occurring in the ablation zone between 1,000 and 1,500 m, in which range the mean near surface air temperature was very close to the melting point.

The spatial distribution of the rate of change in annual albedo shown in Fig. 3.17 indicates that most albedo decreases during the last decade occurred in

Table 3.10 Surface albedo changes at different elevations

Elevation (m)	Annual change rate	
	1981–2000	2000–2012
≤500	0.0004	−0.0006
501 ~ 1000	0.0001	−0.0035**
1001 ~ 1500	−0.0002	−0.0059***
1501 ~ 2000	0.0004	−0.0031**
2001 ~ 2500	0.0005*	−0.0024**
2501 ~ 3000	0.0005**	−0.0015*
>3000	0.0006**	−0.0001

Significance level—*: 90 %; **: 95 %; ***: 99 %; others are not statistically significant at the 90 % level

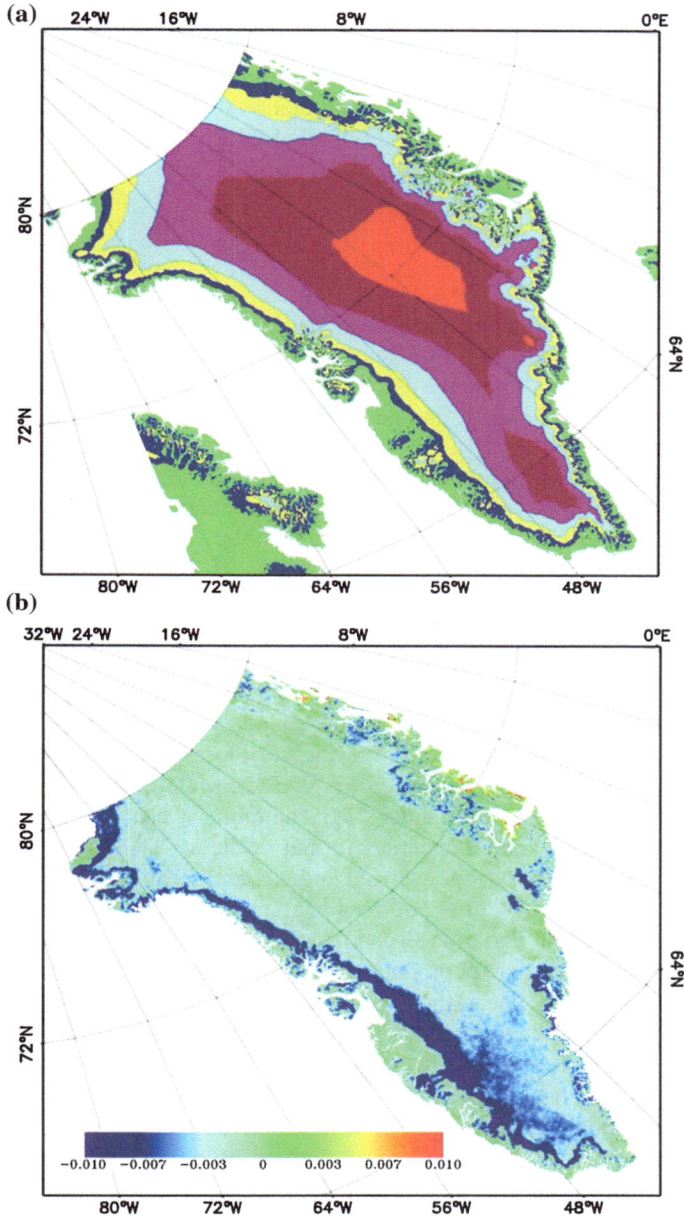

Fig. 3.17 a Digital elevation model of Greenland from USGS GMTED data divided into eight levels: sea level and below (*white*), ≤500 m (*green*), 501–1000 m (*blue*), 1001–1500 m (*yellow*), 1501–2000 m (*cyan*), 2001–2500 m (*magenta*), 2501–3000 m (*maroon*), and above 3000 m (*red*); **b** Annual July albedo change rate over Greenland from the GLASS products for 2000–2012

south-western Greenland, which may be the results of more anticyclonic conditions in Greenland and warmer air advected in this area (Fettweis et al. 2013). Snow melting and the associated albedo decrease may explain the increase in summer modified normalized difference vegetation index (mNDVI) observed over the west coast of Greenland between 1982 and 2010.

3.5 Summary

To facilitate climate modeling and global change studies, the GLASS land surface shortwave albedo product has been generated by newly developed algorithms that are different from the commonly used algorithms in other albedo products. Consequently, it is very important and necessary to assess the quality of the GLASS albedo product.

The GLASS albedo product has no spatial gaps over the global land surface and offers a continuous time series from 1981 to 2010, except in the latter part of 1994, when no good-quality AVHRR data are available. By direct comparison with ground measurements at a set of homogeneous FLUXNET sites, the GLASS albedo product shows a reasonable consistency with the magnitude and trend of the ground measurements, with a bias of less than 0.001 and a RMSE of less than 0.05 for clear days. The temporal resolution is also enhanced; in several examples, a quick change in the ground measurements can be better captured by the daily GLASS albedo product than by the 8-day product. Cross-comparison shows that the GLASS product has very similar values to those from the 1 km-resolution MODIS MCD43B3 product generated by NASA's MODLAND team, especially for data marked by a "good" quality flag. Therefore, it can be expected that the GLASS product has a similar accuracy to that of the MCD43 product, which has been widely validated and acknowledged in the community. On the other hand, the standard MCD43A3 product has 500 m spatial resolution and offers the advantage of capturing the spatial variation of land surface albedo.

The QC flag in the GLASS product gives a valuable indication of good-quality data and those that are uncertain. It is apparent that more than half the data are flagged as "good" or "acceptable." The part of the GLASS product that is most likely to be subject to error is the AVHRR-derived product. There are almost no adequate validation data before 2000; the only approach is to resort to an indirect evaluation method by comparing the AVHRR-derived product with a theoretically more accurate product, i.e., the MODIS-derived product. Good consistency has been found between these two parts of the product.

However, there are possible deficiencies in the current version of the GLASS albedo product, which may arise either from the algorithm or from the input remote sensing data. The main inversion algorithms, i.e., AB1 and AB2, are regression algorithms that are simple and rely on the representativeness of the training dataset. Although these algorithms are optimized in a statistical sense, they certainly cannot account for all situations in the complex real world. There will be remnant errors in the atmospheric correction, anisotropy correction, or band conversion processes. These remnant errors can be partly suppressed by the STF algorithm. However, STF is not guaranteed to perform well in all cases because it is also a statistically based algorithm. STF also introduces side effects, such as oversmoothing that decreases the actual temporal resolution of the filtered results. The a priori database, which is critical in the STF algorithm, is derived from the MCD43B3 product; therefore, this database inherits the limitations of the MCD43B3 product, such as insufficient sampling in tropical rain forest areas.

The GLASS albedo product can be improved in many aspects, for example, remote sensing data from a new sensor should be used, and an inversion algorithm based on multisource data needs to be developed. This work has already been started by scientists. The evaluation and validation of GLASS or any other global albedo product is an ongoing task for all data users. All the validation results, comments, and suggestions will be valuable materials to help improve the GLASS products in the future.

References

Abdalati W, Steffen K (1997) Snowmelt on the Greenland ice sheet as derived from passive microwave satellite data. J Clim 10:165–175

Bacour C, Breon F (2005) Variability of biome reflectance directional signatures as seen by POLDER. Remote Sens Environ 98:80–95

Baldocchi D, Falge E, Gu L, Olson R, Hollinger D, Running S, Anthoni P, Bernhofer C, Davis K, Evans R (2001) FLUXNET: a new tool to study the temporal and spatial variability of ecosystem-scale carbon dioxide, water vapor, and energy flux densities. Bull Am Meteorol Soc 82:2415–2434

Baret F, Morissette JT, Fernandes RA, Champeaux JL, Myneni RB, Chen J, Plummer S, Weiss M, Bacour C, Garrigues S, Nickeson JE (2006) Evaluation of the representativeness of networks of sites for the global validation and intercomparison of land biophysical products: proposition of the CEOS-BELMANIP. IEEE Trans Geosci Remote Sens 44:1794–1803

Box JE, Fettweis X, Stroeve JC, Tedesco M, Hall DK, Steffen K (2012) Greenland ice sheet albedo feedback: thermodynamics and atmospheric drivers. Cryosphere 6:821–839

Bromwich DH, Chen QS, Li YF, Cullather RI (1999) Precipitation over Greenland and its relation to the North Atlantic oscillation. J Geophys Research-Atmos 104:22103–22115

Cescatti A, Marcolla B, Vannan SKS, Pan JY, Roman MO, Yang X, Ciais P, Cook RB, Law BE, Matteucci G, Migliavacca M, Moors E, Richardson AD, Seufert G, Schaaf CB (2012) Intercomparison of MODIS albedo retrievals and in situ measurements across the global FLUXNET network. Remote Sens Environ 121:323–334

Chapin FS, Sturm M, Serreze MC, McFadden JP, Key JR, Lloyd AH, McGuire AD, Rupp TS, Lynch AH, Schimel JP, Beringer J, Chapman WL, Epstein HE, Euskirchen ES, Hinzman LD, Jia G, Ping CL, Tape KD, Thompson CDC, Walker DA, Welker JM (2005) Role of land-surface changes in Arctic summer warming. Science 310:657–660

Comiso JC (2003) Warming trends in the Arctic from clear sky satellite observations. J Clim 16:3498–3510

Cui Y, Mitomi Y, Takamura T (2009) An empirical anisotropy correction model for estimating land surface albedo for radiation budget studies. Remote Sens Environ 113:24–39

Danielson JJ, Gesch DB (2011) Global multi-resolution terrain elevation data 2010. In: U.S. Department of the Interior and U.S. Geological Survey

Dickinson RE (1983) Land surface processes and climate surface albedos and energy-balance. Adv Geophys 25:305–353

Diner DJ, Martonchik JV, Borel C, Gerstl S, Gordon HR, Knyazikhin Y, Myneni R, Pinty B, Verstraete MM (2008) Multi-angle imaging spectroradiometer (MISR) level 2 surface retrieval algorithm theoretical basis (version E). Jet Propulsion Laboratory, Pasadena

Fang H, Liang S, Kim H-Y, Townshend JR, Schaaf CL, Strahler AH, Dickinson RE (2007) Developing a spatially continuous 1 km surface albedo data set over North America from Terra MODIS products. J Geophys Research-Atmos 112:D20206. doi:20210.21029/22006JD008377

Fettweis X, Hanna E, Lang C, Belleflamme A, Erpicum M, Gallée H (2013) Brief communication "Important role of the mid-tropospheric atmospheric circulation in the recent surface melt increase over the Greenland ice sheet". Cryosphere 7:241–248

Gao F, Schaaf C, Strahler A, Roesch A, Lucht W, Dickinson R (2005) MODIS bidirectional reflectance distribution function and albedo climate modeling grid products and the variability of albedo for major global vegetation types. J Geophys Res 110:D01104

Geiger B, Roujean J, Carrer D, Meurey C (2005) Product user manual (PUM) land surface albedo. LSA SAF internal documents

Geiger B Samain O (2004) Albedo determination, algorithm theoretical basis document of the CYCLOPES project. In: Météo-France/CNRM, p 20

Govaerts Y, Lattanzio A, Taberner M, Pinty B (2008) Generating global surface albedo products from multiple geostationary satellites. Remote Sens Environ 112:2804–2816

He T, Liang S, Yu Y, Wang D, Gao F, Liu Q, Greenland surface albedo changes in July 1981–2012 from satellite observations, Environmental Research Letters, (in press)

Hu BX, Lucht W, Strahler AH, Schaaf CB, Smith M (2000) Surface albedos and angle-corrected NDVI from AVHRR observations of South America. Remote Sens Environ 71:119–132

Kendall MG (1976) Rank correlation methods. 4th edn., Griffin, London

Leroy M, Deuzé J, Bréon F, Hautecoeur O, Herman M, Buriez J, Tanré D, Bouffies S, Chazette P, Roujean J (1997) Retrieval of atmospheric properties and surface bidirectional reflectances over land from POLDER/ADEOS. J Geophys Res 102:17023–17037

Li XW, Gao F, Wang JD, Strahler A (2001) A priori knowledge accumulation and its application to linear BRDF model inversion. J Geophys Research-Atmos 106:11925–11935

Liang S (2001) Narrowband to broadband conversions of land surface albedo I: algorithms. Remote Sens Environ 76:213–238

Liang S (2003) A direct algorithm for estimating land surface broadband albedos from MODIS imagery. IEEE Trans Geosci Remote Sens 41:136–145

Liang S (2004) Quantitative remote sensing of land surface. Wiley, New Jersey

Liang S (ed) (2008) Advances in land remote sensing: system, modeling, inversion and application. Springer, Berlin

Liang S, Li X, Wang J (eds) (2012) Advanced remote sensing: terrestrial information extraction and applications. Academic Press, Oxford

Liang S, Strahler A, Walthall C (1999) Retrieval of land surface albedo from satellite observations: a simulation study. J Appl Meteorol 38:712–725

Liang S, Stroeve J, Box J (2005) Mapping daily snow/ice shortwave broadband albedo from moderate resolution imaging spectroradiometer (MODIS): the improved direct retrieval algorithm and validation with Greenland in situ measurement. J Geophys Res 110:D10109

Liang S, Wang K, Zhang X, Wild M (2010) Review on estimation of land surface radiation and energy budgets from ground measurement, remote sensing and model simulations. IEEE J Spec Top Appl Earth Obs Remote Sens 3:225–240

Liang S, He T, Zhang X, Cheng J, Wang D (2013a) Remote sensing of earth surface radiation budget, in remote sensing of land surface turbulent fluxes and soil surface moisture content: state of the art. In: Petropoulos GP (ed), CRC Press, Boca raton, pp 125–165

Liang S, Zhao X, Yuan W, Liu S, Cheng X, Xiao Z, Zhang X, Liu Q, Cheng J, Tang H, Qu YH, Bo Y, Qu Y, Ren H, Yu K, Townshend J (2013b) A long-term global land surface satellite (GLASS) data-set for environmental studies. Int J Digit Earth. doi:10.1080/17538947.17532013.17805262

Liu NF, Liu Q, Wang LZ, Liang SL, Wen JG, Qu Y, Liu SH (2013a) A statistics-based temporal filter algorithm to map spatiotemporally continuous shortwave albedo from MODIS data. Hydrol Earth Syst Sci 17:2121–2129

Liu Q, Wang L, Qu Y, Liu N, Liu S, Tang H, Liang S (2013b) Preliminary evaluation of the long-term GLASS albedo product. Int J Digit Earth. doi:10.1080/17538947.17532013.17804601

Liu Q, Wen JG, Qu Y, He T, Zhang XT (2012) Broadband albedo. In: Liang S, Li X, Wang J (eds.) Advanced remote sensing: terrestrial information extraction and applications Academic Press, Oxford, pp 173–230

Long CN, Gaustad KL (2004) The shortwave (SW) clear-sky detection and fitting algorithm: algorithm operational details and explanations. In: Atmospheric radiation measurement program technical report, 26 pp

Lucht W, Schaaf C, Strahler A (2002) An algorithm for the retrieval of albedo from space using semiempirical BRDF models. IEEE Trans Geosci Remote Sens 38:977–998

Lucht W, Schaaf CB, Strahler AH (2000) An algorithm for the retrieval of albedo from space using semiempirical BRDF models. IEEE Trans Geosci Remote Sens 38:977–998

Maignan F, Bréon F, Lacaze R (2004) Bidirectional reflectance of earth targets: evaluation of analytical models using a large set of spaceborne measurements with emphasis on the hot spot. Remote Sens Environ 90:210–220

Mann HB (1945) Nonparametric tests against trend. Econometrica 13:245–259

Mason P (2005). Implementation plan for the global observing systems for climate in support of the UNFCCC. In: 21st international conference on interactive information processing systems for meteorology, oceanography, and hydrology. San Diego

Moody EG, King MD, Platnick S, Schaaf CB, Feng G (2005) Spatially complete global spectral surface albedos: value-added datasets derived from Terra MODIS land products. IEEE Trans Geosci Remote Sens 43:144–158

Mote TL (2007) Greenland surface melt trends 1973-2007: evidence of a large increase in 2007. Geophys Res Lett, 34:L22507

Muller J-P, Preusker R, Fischer J, Zuhlke M, Brockmann C, Regner P (2007) ALBEDOMAP: MERIS land surface albedo retrieval using data fusion with MODIS BRDF and its validation using contemporaneous EO and in situ data products. In: Geoscience and remote sensing symposium, 2007. IGARSS 2007. IEEE International, IEEE, pp 2404–2407

Nghiem SV, Hall DK, Mote TL, Tedesco M, Albert MR, Keegan K, Shuman CA, DiGirolamo NE, Neumann G (2012) The extreme melt across the Greenland ice sheet in 2012. Geophys Res Lett 39:L20502

Pedelty J, Devadiga S, Masuoka E, Brown M, Pinzon J, Tucker C, Roy D, Ju JC, Vermote E, Prince S, Nagol J, Justice C, Schaaf C, Liu JC, Privette J, Pinheiro A (2007) Generating a long-term land data record from the AVHRR and MODIS instruments. In: Ieee international geoscience and remote sensing symposium, Ieee, New York, pp 1021–1024

Pinty B, Roveda F, Verstraete M, Gobron N, Govaerts Y, Martonchik J, Diner D, Kahn R (2000) Surface albedo retrieval from meteosat 1 theory. J Geophys Res 105:18099–18112

Qin W, Herman J, Ahmad Z (2001) A fast, accurate algorithm to account for non-Lambertian surface effects on TOA radiance. J Geophys Res 106:22671–22684

Qu Y, Liu Q, Liang SL, Wang LZ, Liu NF, Liu SH (2013) Direct-estimation algorithm for mapping daily land-surface broadband albedo from MODIS data. IEEE Trans Geosci Remote Sens. doi:10.1109/TGRS.2013.2245670

Rahman H, Pinty B, Verstraete M (1993) Coupled surface-atmosphere reflectance (CSAR) model 2. semiempirical surface model usable with NOAA advanced very high resolution radiometer data. J Geophys Res 98:20791–20801

Roujean JL, Leroy M, Deschamps PY (1992) A bidirectional reflectance model of the earth's surface for the correction of remote sensing data. J Geophys Res 97:20455–20468

Rutan D, Charlock T, Rose F, Kato S, Zentz S, Coleman L (2006) Global surface albedo from CERES/TERRA surface and atmospheric radiation budget (SARB) data product. In: Proceedings of 12th conference on atmospheric radiation (AMS). Madison

Saunders RW (1990) The determination of broad band surface albedo from AVHRR visible and near-infrared radiances. Int J Remote Sens 11:49–67

Schaaf C, Gao F, Strahler A, Lucht W, Li X, Tsung T, Strugll N, Zhang X, Jin Y, Muller P, Lewis P, Barnsley M, Hobson P, Disney M, Roberts G, Dunderdale M, Doll C, d'Entremont R, Hu B, Liang S, Privette J, Roy D (2002) First operational BRDF, albedo nadir reflectance products from MODIS. Remote Sens Environ 83:135–148

Schaaf C, Martonchik J, Pinty B, Govaerts Y, Gao F, Lattanzio A, Liu J, Strahler A, Taberner M (2008) Retrieval of surface albedo from satellite sensors. In: Liang S (ed) Advances in land remote sensing: system, modelling, inversion and application, Springer, Heidelberg, pp 219–243

Strahler A, Muller J, Lucht W, Schaaf C, Tsang T, Gao F, Li X, Lewis P, Barnsley M (1999) MODIS BRDF/albedo product: algorithm theoretical basis document version 5.0. MODIS documentation

Stroeve J (2001) Assessment of Greenland albedo variability from the advanced very high resolution radiometer polar pathfinder data set. J Geophys Research-Atmos 106:33989–34006

Stroeve J, Box J, Gao F, Liang S, Nolin A, Schaaf C (2005) Accuracy assessment of the MODIS 16-day albedo product for snow: comparisons with Greenland in situ measurements. Remote Sens Environ 94:46–60

Strugnell NC, Lucht W, Schaaf C (2001) A global albedo data set derived from AVHRR data for use in climate simulations. Geophys Res Lett 28:191–194

Trishchenko AP, Luo Y, Khlopenkov KV, Wang S (2008) A method to derive the multispectral surface albedo consistent with MODIS from historical AVHRR and VGT satellite data. J Appl Meteorol Climatol 47:1199–1221

van Leeuwen W, Roujean J (2002) Land surface albedo from the synergistic use of polar (EPS) and geo-stationary (MSG) observing systems: an assessment of physical uncertainties. Remote Sens Environ 81:273–289

Vermote E, Tanré D, Deuzé J, Herman M, Morcrette J (1997) Second simulation of the satellite signal in the solar spectrum(6S), 6S User guide version 3

Vermote EF, El Saleous NZ, Justice CO (2002) Atmospheric correction of MODIS data in the visible to middle infrared: first results. Remote Sens Environ 83:97–111

Wanner W, Li X, Strahler A (1995) On the derivation of kernels for kernel-driven models of bidirectional reflectance. J Geophys Res 100:21077–21090

Weiss M, Baret F, Leroy M, Begue A, Hautecoeur O, Santer R (1999) Hemispherical reflectance and albedo estimates from the accumulation of across-track sun-synchronous satellite data. J Geophys Research-Atmos 104:22221–22232

Zhang X, Liang S, Wang K, Li L, Gui S (2010) Analysis of global land surface shortwave broadband albedo from multiple data sources. IEEE J Sel Top Appl Earth Obs Remote Sens 3:296–305

Chapter 4
Longwave Emissivity

Abstract Land surface broadband emissivity (BBE) is a key parameter for estimating surface radiation budget, but it is treated crudely in land surface models because of a lack of global-scale observational BBE data. This chapter describes the development of new algorithms for estimating global land surface BBE and production of the global 8-day, 1 and 5 km land surface BBE product from 1981 to 2010. Section 4.1 reviews the need for BBE by land surface models and the drawbacks of existing emissivity and BBE datasets; Sect. 4.2 introduces the GLASS BBE algorithms; Sect. 4.3 describes the characteristics, quality control, and validation of the GLASS BBE product; Sect. 4.4 provides a preliminary analysis of this new BBE product, and Sect. 4.5 provides a short summary.

Keywords Broadband emissivity · Surface radiation budget · Land surface model · Albedo · NDVI · Soil taxonomy · Remote sensing

4.1 Background

Land surface thermal infrared broadband emissivity (BBE) is an essential parameter for simulating surface energy budgets (Cheng and Liang 2013b; Liang et al. 2010; Pequignot et al. 2008; Sellers et al. 1997). Given the lack of effective BBE datasets, constant emissivity assumption or simple parameterization schemes are used in current land surface and general circulation models (Bonan et al. 2002; Jin and Liang 2006; Zhou et al. 2003). For example, the National Center for Atmospheric Research (NCAR) Community Land Model Version 2 (CLM2) calculates canopy emissivity from LAI, and sets soil and snow emissivities as 0.96 and 0.97 (Bonan et al. 2002).

A number of researchers have generated several BBE data sets on a global or regional scale using different methods. Wilber et al. established a global BBE

S. Liang et al., *Global LAnd Surface Satellite (GLASS) Products*,
SpringerBriefs in Earth Sciences, DOI: 10.1007/978-3-319-02588-9_4,
© The Author(s) 2014

(5–100 μm) with a $10' \times 10'$ spatial resolution for satellite retrieval of long-wave radiation by assigning constant emissivity values to International Geosphere–Biosphere Program (IGBP) surface types (Wilber et al. 1999). Ogawa et al. mapped the monthly global BBE (8.0–13.5 μm) using the Moderate-resolution Imaging Spectroradiometer (MODIS) narrowband emissivity product (approximately 5 km) and a North African BBE (8.0–13.5 μm) using the Advanced Spaceborne Thermal Emission and Reflectance Radiometer (ASTER) narrowband emissivity product (90 m) (Ogawa and Schmugge 2004; Ogawa et al. 2008).

Two drawbacks were observed. (1) Either the spatial or temporal resolution of these products is limited. BBE with finer spatial and temporal resolutions would be useful for local-scale studies of surface energy balance and for validating coarse resolution data, thus facilitating better understanding of land–atmosphere interactions (Liang et al. 2010; Ogawa and Schmugge 2004). (2) These products cannot capture variations in natural land surface BBE. In Wilber et al., seasonal variation was neglected, particularly for vegetated surfaces, because each IGBP surface type had only one emissivity over the entire growth stage. The MODIS emissivity product cannot reflect the effect of vegetation cover changes on the emissivity of vegetated surfaces (Wang and Liang 2009a). A number of pioneering studies have demonstrated the importance of satellite-derived realistic BBE in improving the results of global climate model simulations (Jin and Liang 2006; Zhou et al. 2003). Therefore, the global land surface BBE products with high spatial and temporal resolutions are needed.

4.2 Algorithms

Converting satellite-derived narrowband emissivity products is a common method for estimating BBE (Tang et al. 2011; Wang et al. 2005). However, retrieving land surface emissivity (LSE) from radiometric measurements of a thermal infrared sensor is inherently an ill-posed problem, to which solutions are likely unstable (Li et al. 1999, 2013; Liang 2001). Numerous temperature and emissivity inversion algorithms have been proposed to solve this problem since the 1980s (Cheng et al. 2008, 2010a, 2011; Dash et al. 2002; Gillespie et al. 1998, 2011; Li and Becker 1993; Liang 2001, 2004; Liu and Xu 1998; Wan and Li 1997; Wang et al. 2008; Yu et al. 2009), however, only a few algorithms have been adapted to produce global LSE products, such as the temperature and emissivity separation (TES) algorithm for ASTER (Gillespie et al. 1998) and the day/night algorithm for MODIS (Wan and Li 1997).

Cheng and Liang examined potential methods for retrieving global land surface BBE with high spatial and temporal resolutions (Cheng and Liang 2013a). The ASTER narrowband emissivity product derived through the TES algorithm is inefficient because of its long revisit time of 16 days. The MODIS narrowband emissivity product retrieved through the day/night algorithm lacks extensive validation and sufficient recognition from the remote sensing community, thus

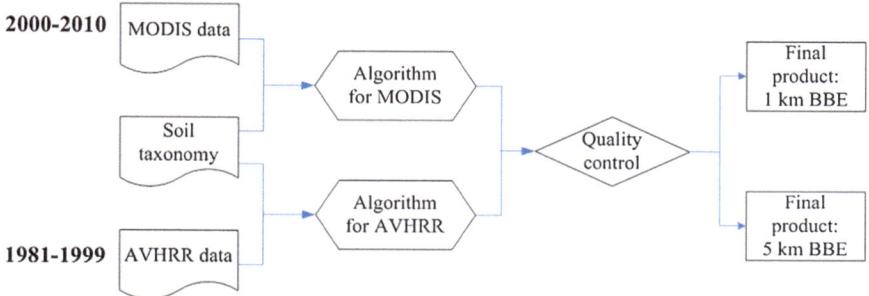

Fig. 4.1 Overview of the GLASS BBE algorithm

making it unsuitable for producing the global land surface BBE products with high spatial and temporal resolutions. A new algorithm must be developed to accomplish this difficult task.

Figure 4.1 shows an overview of the GLASS BBE algorithm. The details of the algorithm are described in the following sections.

4.2.1 Determining the Optimal Broadband Emissivity Spectral Range

The spectral domain needed for calculating surface longwave net radiation is the total longwave range, whereas existing thermal infrared (TIR) sensors can only provide several discrete narrowband emissivities within the spectral range of 3–14 μm. Many studies have been carried out with the aim of estimating BBE in different spectral domains (e.g., 3–14 μm, 8–12 μm, 8–13.5 μm, 3–∞ μm, and 0–∞ μm) by remote sensing (Ogawa and Schmugge 2004; Tang et al. 2011; Wang et al. 2005). Ogawa and Schmugge (2004) compared the accuracy of BBE defined for three spectral domains (3–14 μm, 8–12 μm, 8–13.5 μm) in estimating surface longwave net radiation; they claimed that 8–13.5 μm is the best spectral domain. However, the actual spectral domain for the emissivity spectra used in these studies is limited to 3–14 μm. The meteorological, hydrological, and agricultural research communities require an accuracy of 5–10 W/m^2 for the satellite-derived surface longwave radiation budget at 3 h to daily temporal resolution and 25–100 km spatial resolution (CEOS and WMO 2000). The acceptable accuracy for instantaneous surface longwave radiation is 20 W/m^2 (Gupta et al. 2004), but the accuracy of satellite-derived instantaneous surface longwave radiation does not meet this goal (Wang and Liang 2009b). The replacement of all-wavelength BBE with remote sensing BBE when calculating surface longwave net radiation might be a potential source of error.

The surface longwave net radiation, L_n, is the difference between the surface upwelling longwave radiation and the surface downward longwave radiation.

$$L_n = \int_{\lambda 1}^{\lambda 2} \varepsilon_\lambda [B(T_s) + \rho_\lambda L_{a\lambda}] \mathrm{d}\lambda - \int_{\lambda 1}^{\lambda 2} L_{a\lambda} \mathrm{d}\lambda \tag{4.1}$$

where ε_λ is the surface spectral emissivity, $B(T_s)$ is the Planck function at temperature T_s, ρ_λ is the directional hemispherical spectral reflectance, and $L_{a\lambda}$ is the surface downward longwave radiation. The lower and upper wavelengths considered for integration are $\lambda 1 = 0$ and $\lambda 2 = \infty$, respectively. Assuming that the surface is under thermodynamic equilibrium and follows Kirchhoff's law, Eq. (4.1) can be expressed as

$$L_n = \int_{\lambda 1}^{\lambda 2} \varepsilon_\lambda [B(T_s) - L_{a\lambda}] \mathrm{d}\lambda \tag{4.2}$$

Assuming that ε_λ is independent of $B(T_s)$ and $L_{a\lambda}$, and neglecting bandpass effect, Eq. (4.2) can be formulated as Eq. (4.3), which is used in land surface models (Bonan et al. 2002). This assumption will affect the accuracy of the L_n calculation. Using the data in Sect. 2.2, the bias and root mean square error (RMSE) obtained by replacing Eq. (4.2) by Eq. (4.3) were 0.55 and 2.31 W/m^2, respectively, for the 2.5–200 µm spectral domain.

$$L_n = \varepsilon_{\mathrm{BB}} \left(\sigma T_s^4 - L_a \right) \tag{4.3}$$

where σ is the Stefan-Boltzmann constant (5.67×10^{-8} W/m^2/K^{-4}). L_a is the wavelength integration of $L_{a\lambda}$, and $\varepsilon_{\mathrm{BB}}$ is the BBE. These latter two quantities are defined in Eqs. (4.4) and (4.5):

$$L_a = \int_{\lambda 1}^{\lambda 2} L_{a\lambda} \mathrm{d}\lambda \tag{4.4}$$

$$\varepsilon_{\mathrm{BB}} = \frac{\int_{\lambda 1}^{\lambda 2} \varepsilon_\lambda B_\lambda(T_s) \mathrm{d}\lambda}{\int_{\lambda 1}^{\lambda 2} B_\lambda(T_s) \mathrm{d}\lambda} \tag{4.5}$$

where $\lambda 1$ and $\lambda 2$ specify the spectral range of the surface longwave radiation. Theoretically, the spectral range of integration in Eq. (4.3) should be 0–∞ µm. However, it is impossible to determine ε_λ and $L_{a\lambda}$ over such a broad spectral range using modern sensors or simulation tools. Currently, the spectral range of 4–100 µm is used to estimate the surface longwave radiation budget (Wang et al. 2009). The 4–100 µm spectral range accounts for 99.5 % of the total radiation of a 300 K blackbody, whereas the 1–200 µm spectral range accounts for 99.92 %. The accuracy of calculating the all-wavelength surface longwave net radiation using the 1–200 μm spectral domain would then be better than that using the 4–100 μm spectral domain. Ignoring the emissivity outside the 1–200 μm spectral range has little impact on the calculation of all-wavelength surface longwave net radiation.

Soil, rock, vegetation, water, and snow/ice are the primary components of land surfaces. One of the main components of soil and rock is minerals. The emissivity spectra of minerals share some common features with those of soil and rock and can replace the emissivity spectra of soil and rock to some extent. Minerals and snow/ice are layered particulate media. Their emissivity spectra can be simulated using modern radiative transfer tools by incorporating scattering parameters (Cheng et al. 2010b; Pitman et al. 2005). Pitman et al. (2005 showed that modern radiative transfer tools can replicate the shape of laboratory-measured 5–50 μm emissivity spectra of quartz with different radii).

Cheng et al. examined the capability of modern radiative transfer models for simulating snow TIR emissivity spectra at 8–13 μm (Cheng et al. 2010b). Their work revealed that both the shape and value of field-measured nadir emissivity spectra for different particle sizes can be reproduced properly. Given the refractive index (optical constant) of pure water, its emissivity spectra can be calculated by the Fresnel equation (Masuda et al. 1988). For minerals and snow/ice, the Mie theory was used to calculate the scattering parameters and to correct for the dense packing of particles (Cheng et al. 2010b). The corrected scatter parameters were used as input to the Hapke model to simulate the nadir emissivity spectra. For pure water, the Fresnel equation was used here to simulate the nadir emissivity spectra. For vegetation, the average emissivity spectra of three live leaves in the ASTER spectral library (http://speclib.jpl.nasa.gov/) was used. L_n values were calculated for different spectral domains, and their accuracy for replacing the all-wavelength L_n was assessed. On the basis of this accuracy, the optimal spectral domain for calculating L_n was identified. Next, the accuracy of BBE was investigated for different spectral domains when calculating L_n for the optimal spectral domain.

The refractive indices used are presented in Table 4.1. In the simulation of scattering parameters for ice and minerals, the spectral range was set to 1–200 μm.

Table 4.1 Spectral range and the source of the refractive index used

Name	Spectral range (μm)	Source
Water	0.2–200	Hale and Querry (1973)
Ice	0.16–10,000	Warren and Brandt (2008)
10 Phyllosilicate minerals	5–100	Glotch et al. (2007)
3 Silicon dioxide	6.7–500	Henning et al. (1999)
Aluminum oxide	7.8–500	Henning et al. (1999)
Iron oxide (Mg/Fe oxides)	0.2–500	Henning et al. (1999)
Silicate (Mg/Fe olivine)	0.2–500	Henning et al. (1999)
2 Silicate (O-rich circumstellar; O-deficient circumstellar)	0.4–10,000	Henning et al. (1999)

If the starting wavelength or ending wavelength of the refractive index fell into the 1–200 μm spectral region, the refractive index was extrapolated to 1–200 μm by the nearest neighbor method (i.e., at wavelengths from 1 μm to the starting wavelength, the refractive index equaled that at the starting wavelength; and at wavelengths from the ending wavelength to 200 μm, the refractive index equaled that at the ending wavelength). The average particle radii of fresh snow, old snow, and minerals were approximately 40, 300, and 300 μm, respectively (Hori et al. 2006; Pitman et al. 2005). The radius for ice was specified as 40 or 300 μm, and the radius for minerals was set to 300 μm.

Figure 4.2 shows the simulated emissivity spectra of snow, water, and some of the minerals. As shown in Fig. 4.2, except for one mineral sample, the spectral contrast between the simulated emissivity spectra at 30–200 μm was very small, whereas the spectral contrast at 1–30 μm was quite large. Obviously, replacing the emissivity in the 14–200 μm spectral range with any one emissivity value is not advisable. The 2,311 atmosphere profiles in the latest Thermodynamic Initial Guess Retrieval (TIGR) database (Chedin et al. 1985) were used to simulate atmospheric downward radiance in the 0.2–1,000 μm spectral range. MODTRAN4 was used to simulate the directional atmospheric downward radiation at five-degree intervals. Assuming that the directional atmospheric downward radiation is independent of the azimuth angle, $L_{a\lambda}$ was calculated by integration.

Calculation of the all-wavelength L_n in real applications is impractical. The accuracy of L_n for the 4–100 μm spectral domain and other spectral domains (e.g., 3–100 μm, 2.5–100 μm, 2.5–200 μm, and 1–200 μm) was investigated for replacing the all-wavelength L_n. All-wavelength L_n and L_n for other spectral domains were calculated using Eq. (4.2). The 21 simulated emissivity spectra were extrapolated to 0.2–1,000 μm by the nearest neighbor method. The vegetation emissivity spectrum was extrapolated to 0.2–1,000 μm by the same method. Surface temperature was specified as T_a, $T_a + 8$, and $T_a + 15$, where T_a is the near-surface air temperature of the TIGR atmosphere profiles. The results are

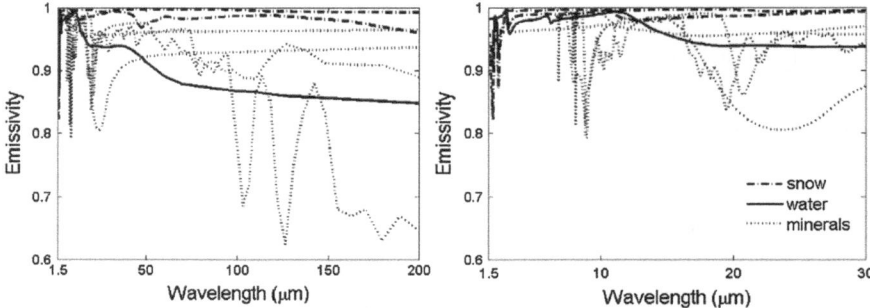

Fig. 4.2 Simulated nadir emissivity spectra for snow, water, and minerals in the 1–200 μm and 1–30 μm spectral ranges, respectively (Cheng et al. 2013). Copyright © 2013 Reproduced with permission of IEEE

presented in Fig. 4.3, when the all-wavelength L_n, was less than -50 W/m^2; the L_n values for the 4–100 μm, 3–100 μm, and 2.5–100 μm spectral domains clearly overestimated the actual value. The L_n values for the 2.5–200 μm and 1–200 μm spectral domains are closely distributed around the 1:1 line. The statistical results are listed in Table 4.2. The bias of spectral domains 4–100 μm, 3–100 μm, and 2.5–100 μm ranged from 3.725 to 4.46 W/m^2, and the RMSE lay between 4.971 and 5.679 W/m^2. The bias and RMSE of the 2.5–200 μm spectral domains were slightly smaller than those of the 1–200 μm spectral domain. The average bias and RMSE of 2.5–200 μm spectral domains were 0.92 and 0.985 W/m^2, respectively, which were 3.052 and 4.148 W/m^{2,} respectively, lower than the values of the

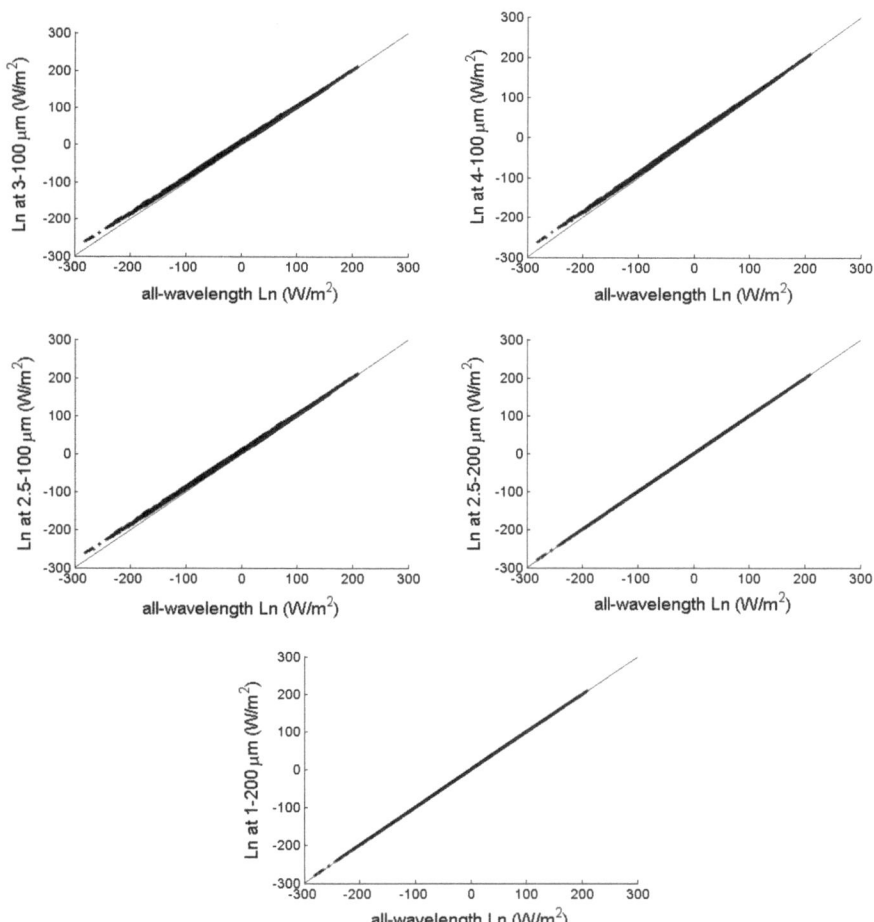

Fig. 4.3 Comparison of all-wavelength L_n and L_n estimated for different spectral domains (Cheng et al. 2013). Copyright © 2013 Reproduced with permission of IEEE

Table 4.2 Bias and RMS of L_n estimates obtained using different spectral domains

Spectral domain (μm)	Bias (W/m^2)			RMSE (W/m^2)		
	T_1	T_2	T_3	T_1	T_2	T_3
4–100	4.212	3.979	3.725	5.387	5.140	4.871
3–100	4.460	4.388	4.322	5.679	5.618	5.562
2.5–100	4.459	4.393	4.336	5.676	5.624	5.578
2.5–200	0.928	0.920	0.912	0.993	0.984	0.977
1–200	0.929	0.921	0.914	0.993	0.986	0.979

$T_1 = T_a$, $T_2 = T_a + 8$, and $T_3 = T_a + 15$

4–100 μm spectral domain. The meteorological, hydrological, and agricultural research communities require an accuracy of 5–10 W/m^2 for the monthly averaged satellite-derived surface longwave radiation budget (CEOS and WMO 2000). Therefore, this improvement will have significant implications.

BBE values were first calculated for the 2.5–200 μm, 3–14 μm, 8–12 μm, 8–13.5 μm, and 8–14 μm spectral domains using Eq. (4.5). BBE is not sensitive to surface temperature (Ogawa and Schmugge 2004), and therefore, the surface temperature was defined as the average Earth's surface temperature (300 °K). Equation (4.3) was then used to calculate L_n for the 2.5–200 μm spectral domain with these five BBE domains mentioned earlier. The spectral range for calculating L_a in Eq. (4.3) was 2.5–200 μm.

The accuracy for each spectral domain is presented in Table 4.3. The 3–14 μm spectral domain had the largest bias. Although it had the lowest RMSE, the difference between its RMSE and those of the other spectral domains was not significant. The 8–12 μm spectral domain had the largest RMSE and had a negative bias with an absolute value lower only than that of the 3–14 μm spectral domain. The 3–14 μm and 8–12 μm spectral domains were found to be inappropriate for calculating L_n for the 2.5–200 μm spectral domain. The 8–13.5 μm spectral domain had the minimum bias and an acceptable RMS. The average bias and RMSE for this domain were 0.001 and 1.184 W/m^2, respectively. The bias of the 8–14 μm spectral domain was larger than that of the 8–13.5 μm spectral domain, and the RMSE of the 8–14 μm spectral domain was slightly less than that of the 8–13.5 μm spectral domain. Therefore, the BBE value of the 8–13.5 μm spectral domain was the best for calculating L_n value of the 2.5–200 μm spectral domain.

Table 4.3 Bias and RMS when estimating L_n at 2.5–200 μm using the BBE for different spectral domains

Spectral domain (μm)	Bias (W/m^2)			RMS (W/m^2)		
	T_1	T_2	T_3	T_1	T_2	T_3
3–14	0.070	0.387	0.687	0.863	0.990	1.292
8–12	−0.046	−0.254	−0.451	1.383	1.585	2.070
8–13.5	0.000	0.001	0.002	0.978	1.120	1.453
8–14	0.010	0.054	0.096	0.911	1.045	1.373

Using these simulation data, the accuracy of the 2.5–200 μm L_n calculated using the 8–13.5 μm BBE to replace all-wavelength L_n was investigated. In this case, the average bias and RMSE were 1.473 and 2.746 W/m^2, respectively.

Previous studies demonstrated that the BBE can be represented by a linear function of satellite narrowband emissivities (Jin and Liang 2006; Tang et al. 2011). Earlier works tended to derive this linear function using one spectral library (the ASTER spectral library) and then to test it using another independent spectral library (the MODIS UCSB spectral library, http://g.icess.ucsb.edu/modis/EMIS/html/em.html). The test results verify the efficacy of the linear function derived from the ASTER library. These two libraries are independent, and therefore, a linear function derived using both libraries would be more representative than one derived by using only the ASTER spectral library. Here, linear functions are presented for converting ASTER and MODIS narrowband emissivities to BBE at 8–13.5 μm using the ASTER spectral library, the MODIS UCSB spectral library, and the soil samples from this study.

A total of 240 spectra were selected from the ASTER spectra, including 186 rock samples, 41 soil samples, 4 vegetation samples, and 9 water/snow/ice samples. A total of 109 spectra were selected from the MODIS UCSB spectral library. In addition, the emissivity spectra of 75 soil samples measured outdoors were used. In total, 424 spectra were used to derive the linear function. ASTER and MODIS narrowband emissivities were calculated using Eq. (4.6):

$$\varepsilon_i = \frac{\int_{\lambda 1}^{\lambda 2} \varepsilon_\lambda f_i(\lambda) B_\lambda(T) d\lambda}{\int_{\lambda 1}^{\lambda 2} f_i(\lambda) B_\lambda(T) d\lambda} \tag{4.6}$$

where $f_i(\lambda)$ is the spectral response function of band i. The BBE for 8–13.5 μm was calculated using Eq. (4.5). Then, the linear functions for ASTER and MODIS were obtained through statistical regression. The conversion equations can be expressed as

$$\varepsilon_{bb\_ast} = 0.197 + 0.025\varepsilon_{10} + 0.057\varepsilon_{11} + 0.237\varepsilon_{12} + 0.333\varepsilon_{13} + 0.146\varepsilon_{14} \tag{4.7}$$

$$\varepsilon_{bb\_mod} = 0.095 + 0.329\varepsilon_{29} + 0.572\varepsilon_{31} \tag{4.8}$$

where ε_{bb_ast} is the ASTER BBE, ε_{10}–ε_{14} are the five ASTER narrowband emissivities, ε_{bb_mod} is the MODIS BBE, and ε_{29} and ε_{31} are the MODIS narrowband emissivities for channels 29 and 31. The R-squared and RMSE for ASTER were 0.983 and 0.005, respectively, while the R-squared and RMSE for MODIS were 0.932 and 0.010, respectively.

The conversion equations were tested using leave-one-out cross validation (Bsaibes et al. 2009). The 423 samples were used to establish the conversion function, and the remaining single sample was used for validation. This process was repeated 424 times to cover the entire set of samples. Each time, one conversion equation was obtained, as well as the corresponding R-squared value and one bias between the predicted BBE and the true BBE. The mean and RMSE of the bias were zero and 0.005, respectively; the R-squared value was 0.983 ± 0.0001 for ASTER. The corresponding values for MODIS were zero, 0.010, and 0.932 ± 0.0005. The coefficients of the conversion function were 0.197 ± 0.0, 0.025 ± 0.0, 0.057 ± 0.0, 0.237 ± 0.0, 0.333 ± 0.0, and 0.146 ± 0.0 for ASTER, and 0.095 ± 0.002, 0.329 ± 0.0, and 0.572 ± 0.002 for MODIS.

4.2.2 Estimating the Longwave Emissivity from MODIS Shortwave Albedos

Based on soil spectra from the Johns Hopkins University Spectral Library, BBE values at 8.0–13.5 μm and the seven corresponding MODIS narrowband black-sky albedos were calculated. A significant linear relationship between BBE and the seven narrowband black-sky albedos was explored. This relationship was initially verified for bare soil (Cheng and Liang 2013a) using spatially and temporally matched ASTER emissivity products and MODIS narrowband albedo products, and then for homogeneous vegetated areas (Ren et al. 2013).

In the algorithm used to generate GLASS BBE from MODIS albedos, land surface was classified into five types according to Normalized Difference Vegetation Index (NDVI) threshold values: water, snow or ice, bare soil ($0 < \text{NDVI} \leq 0.156$), vegetated area ($\text{NDVI} > 0.156$), and transition zone ($0.1 < \text{NDVI} < 0.2$). Areas of overlapping bare soil and transition zone as well as of transition zone and vegetated area were noted. BBEs of water and snow or ice were set to 0.985 based on a combination of BBE calculated from the emissivity spectrum in the ASTER spectral library (http://spclib.jpl.nasa.gov), the MODIS University of California Santa Barbara Emissivity Library (http://www.icess.ucsb.edu/modis/EMIS/html/em.html), and BBE values simulated using radiative transfer models (Cheng et al. 2010b). BBEs of bare soil, vegetated areas, and transition zones were formulated as the linear function of seven MODIS narrowband black-sky albedos. When the NDVI was less than 0.1 or greater than 0.2, the equations for bare soil or vegetated areas were used to calculate individual BBE values. In areas of overlapping bare soil and transition zone ($0.1 < \text{NDVI} \leq 0.156$), the BBE was assumed to be the average values calculated using formulas for bare soil and transition zones. By contrast, BBE for areas of overlapping transition zone and vegetated area ($0.156 < \text{NDVI} < 0.2$) was taken as the average of values calculated using equations for transition zones and vegetated areas.

The scheme for retrieving global land surface BBE for vegetated area for bare soil and transition zone is shown in Fig. 4.4. The satellite data included are (1) the MODIS albedo products (MCD43B3 and MCD43B2), (2) the MODIS normalization vegetation index (NDVI) product (MOD13A2), and (3) the ASTER emissivity product (AST05). The temporal and spatial resolutions of the MCD43B3 product are 8 days and 1 km, respectively. There are two types of albedo: white-sky (diffuse) albedo and black-sky (direct) albedo. The seven narrow band black-sky albedos are

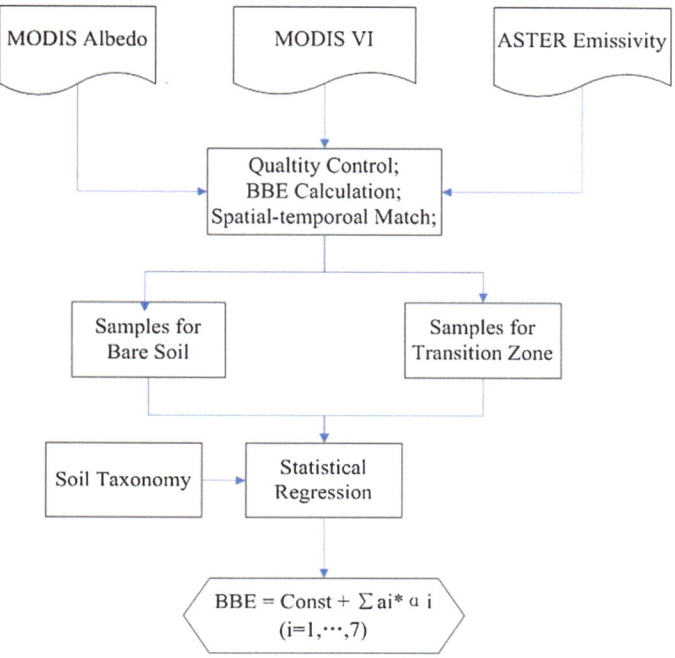

Fig. 4.4 Flowchart for establishing the relationship between the MODIS albedos and ASTER BBE for bare soil and transition zone

used here. The MODIS NDVI was used to identify different land cover types such as bare soils, transition areas, and vegetated areas. The temporal and spatial resolutions of the NDVI are 16 days and 1 km, respectively.

The required auxiliary data are the soil taxonomy, which was used to identify different soil orders. The soil taxonomy map was downloaded from the USDA NRCS (http://soils.usda.gov/use/worldsoils/mapindex/order.html), which is based on a reclassification of the 1994 FAO-UNESCO soil map of the world combined with a soil climate map. The map's spatial resolution is approximately 0.0333°, with 5,400 × 10,800 pixels.

As shown in Fig. 4.5, there are 12 soil orders in the map. According to the soil taxonomy, the study areas were selected and the samples for training and testing the neural network for bare soils were extracted. To extract as many bare soil pixels as possible, an attempt was made to select relatively large and homogeneous areas as the study area for each soil order. The selected study areas are shown in Fig. 4.5. For the soil orders of histosols and spodosols, several places and time intervals were examined, but bare soil pixels were not found. Bare soil pixels were difficult to locate in high latitudes and equatorial zones. Table 4.4 presents the geographical locations of the selected study areas and the time intervals of the data used.

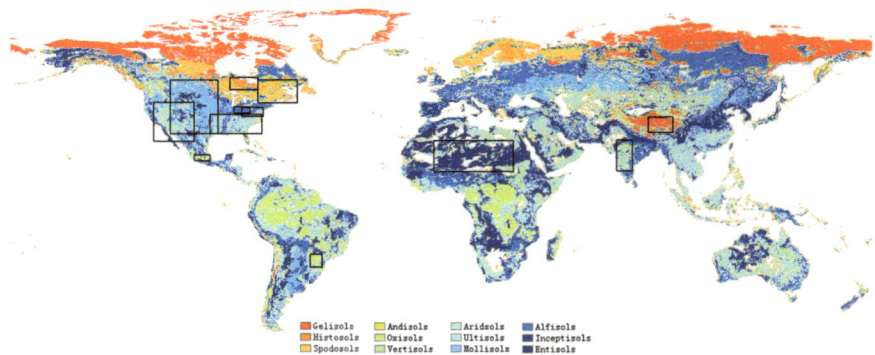

Fig. 4.5 USDA NRCS soil taxonomy and selected experimental areas. Other classes such as rocky land, shifting sand, and ice/glaciers are not included in the map

Table 4.4 Geographic locations of selected experimental areas and time intervals of data used

Soil order	Geographic location	Time interval
Alfisols	Latitude: 38.67°–41.77°	2009.01–2009.03
	Longitude: −86.7° to −80.03°	(2007.01–2007.03)
Andisols	Latitude: 18.3°–20.7°	2009.02–2009.03
	Longitude: −103.3° to −96.7°	(2008.01–2008.03)
Aridisols	Latitude: 26.7°–43.4°	2009.01–2009.02
	Longitude: −120° to −103.3°	(2008.01–2008.03)
Entisols	Latitude: 13.3°–26.7°	2009.02–2009.02
	Longitude: −3.3° to 30°	(2008.01–2008.02)
Gelisols	Latitude: 30.03°–36.7°	2008.01–2008.01
	Longitude: 86.63°–96.63°	(2009.01–2009.04)
Inceptisols	Latitude: 30.03°–36.7°	2008.01–2008.01
	Longitude: 86.63°–96.63°	(2009.01–2009.04)
Mollisols	Latitude: 30°–53.4°	2009.03–2009.03
	Longitude: −113.3° to −93.4°	(2008.01–2008.03)
Oxisols	Latitude: −27.5° to −22°	2008.07–2008.09
	Longitude: −55° to −50°	
Ultisols	Latitude: 30°–38.4°	2008.10–2009.02
	Longitude: −96.7° to −75°	(2007.01–2007.04)
Vertisols	Latitude: 13.3°–26.7°	2009.02–2009.03
	Longitude: 73.3°–80°	(2008.01–2008.03)

The time intervals in parentheses are for the test data

It is desirable to derive one equation using all the extracted BBE-albedo pairs for two reasons: (1) the spatial resolution of the soil taxonomy is much coarser than that of the satellite data used; (2) the accuracy of the soil taxonomy is relatively low. One single equation was derived using all the extracted 46,423 BBE-albedo pairs for bare soil. However, it did not work well for all soil orders combined. Therefore, two equations were derived, i.e., one for andisols alone, and

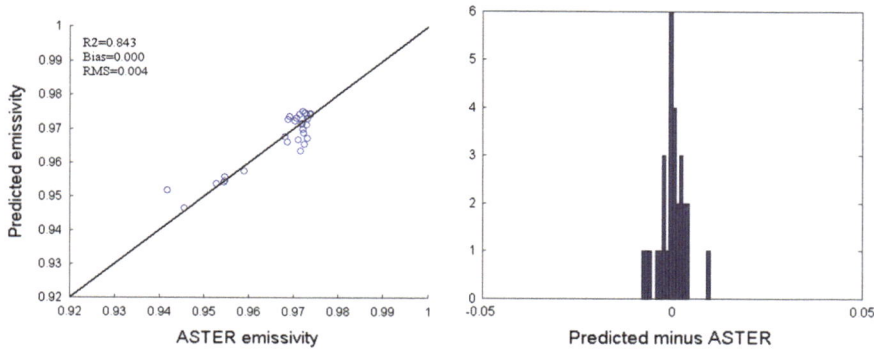

Fig. 4.6 Comparison of ASTER BBE and that predicted by Eq. (4.9) for andisols as well as the histogram of the bias

another for the remaining nine soil orders. Each equation and the coefficient for each variable were significantly below the confidence level of $P < 0.05$. The equations can be expressed as follows:

$$\varepsilon_{BB\_s1} = 0.963 + 0.643a_1 - 1.011a_3 - 0.137a_7 \tag{4.9}$$

$$\varepsilon_{BB\_s2} = 0.954 - 0.945a_1 + 0.518a_2 + 0.743a_3 - 0.154a_4 + 0.019a_5 + 0.095a_6 \\ - 0.117a_7$$

$$\tag{4.10}$$

where ε_{BB_s1} and ε_{BB_s2} are the BBE for andisols and the remaining nine soil orders, respectively.

Figure 4.6 shows the scatter plot for the BBE values predicted using Eq. (4.9) versus the ASTER BBE for andisols and the histogram of the bias. The correlation was 0.843. The bias was centered on zero and distributed in a narrow band, but its value was very small and it can be neglected, while the RMSE was 0.004. Figure 4.7 shows the scatter plot for the BBE values predicted using Eq. (4.10) versus the ASTER BBE for the nine soil orders and the histogram of the bias. The correlation was 0.705. The bias was centered on zero and distributed in a narrow band, but its value was very small and it can be neglected, while the RMSE was 0.011. Table 4.5 shows the bias and RMSE for each of the nine soil orders. The absolute bias and RMSE were less than 0.008 and 0.017, respectively. For other soil orders in the taxonomy, Eq. (4.10) was used to calculate their BBE values.

The performance of the developed algorithm was evaluated by comparing it with the ASTER BBE because of the complexity of the direct validation of emissivity derived from remote sensing data. The geographical location of the test area was the same as that of the experimental area because it was difficult to find bare soil pixels in other locations, although the time intervals were different for the test data (Table 4.4). The BBE-albedo pairs were extracted for all soil orders except oxisols and the derived equations (Eqs. 4.9 and 4.10) were evaluated. For

Fig. 4.7 Comparison of
ASTER BBE and that
predicted by Eq. (4.10) for
the remaining nine soil orders
as well as the histogram of
the bias

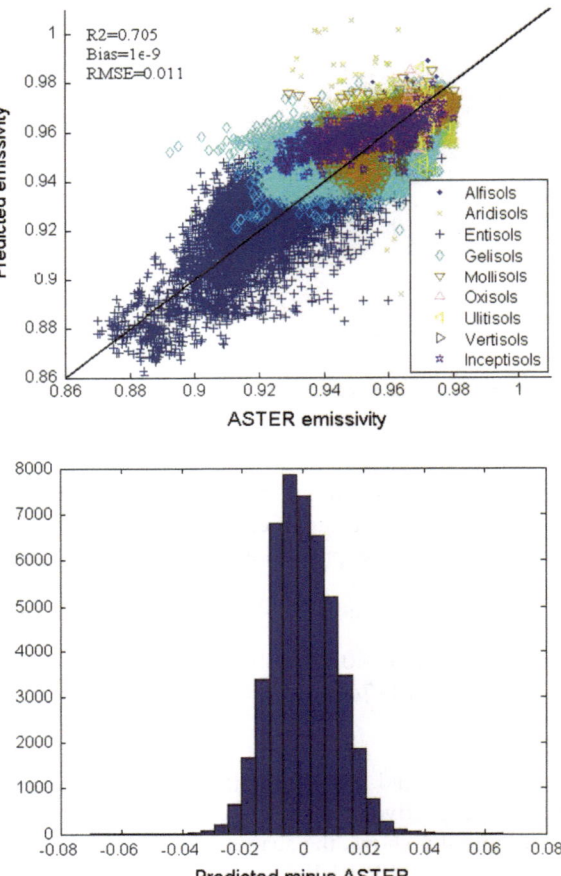

Table 4.5 Bias and RMS for
each of the nine soil orders

Soil order	Bias	RMS
Alfisols	0.003	0.010
Aridisols	−0.002	0.010
Entisols	0.003	0.012
Gelisols	0.000	0.011
Inceptisols	0.007	0.012
Mollisols	−0.003	0.007
Oxisols	−0.008	0.017
Ultisols	−0.008	0.012
Vertisols	−0.005	0.009

Fig. 4.8 Difference histograms of ASTER BBE and that retrieved by the equation for andisols

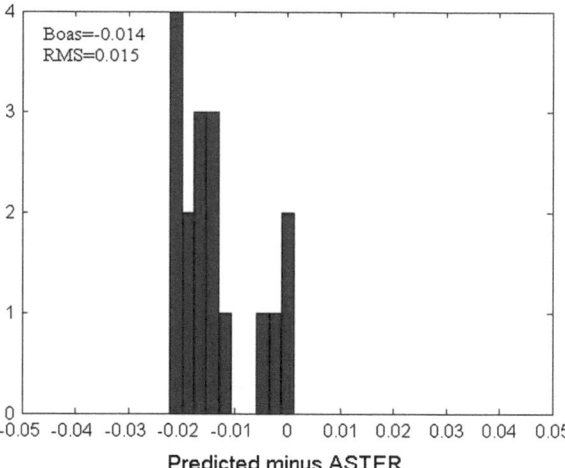

Fig. 4.9 Difference histograms of ASTER BBE and that retrieved by the equation for the remaining nine soil orders

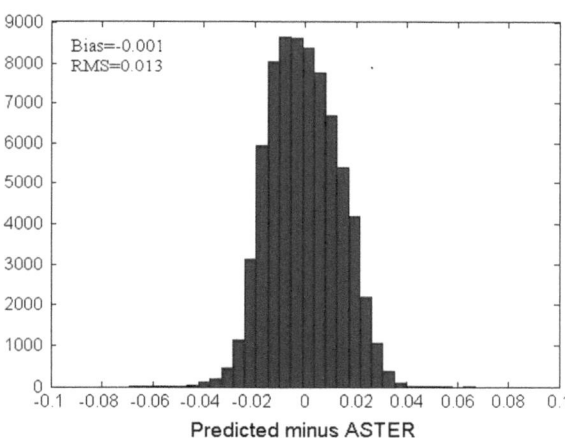

oxisols, over 600 ASTER scenes from 2006 to 2010 were downloaded, but no pixels were found for bare soil and transition zones. Figures 4.8 and 4.9 show the results of the comparison. For andisols, the bias was as high as -0.014, and the RMSE was 0.015. The number of pixels used to establish and test Eq. (4.9) was quite limited, which might have made Eq. (4.9) more unstable and insufficiently representative. The test results may also have been affected by this problem. The BBE predicted by Eq. (4.10) was in good agreement with the ASTER BBE, with a bias and RMSE of -0.001 and 0.013, respectively.

In sparsely vegetated areas, it was hard to determine whether the pixels relate to bare soil or a vegetated surface. The BBE of such pixels was calculated using the equation for bare soil or a vegetated surface. Therefore, the variation in the BBE for these pixels was larger than the actual variation. This led to step discontinuities

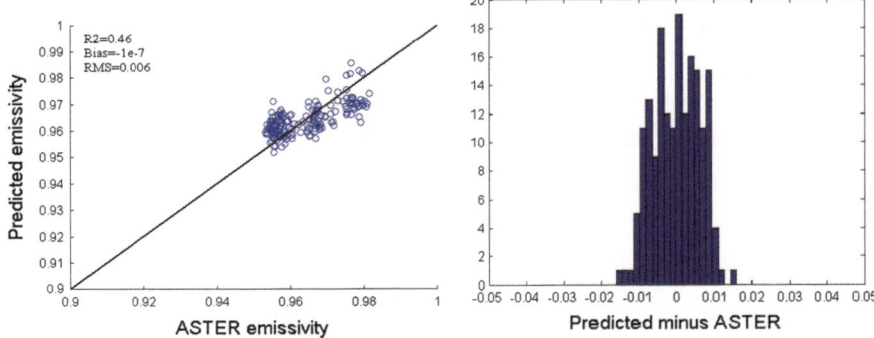

Fig. 4.10 Comparison of ASTER BBE and that predicted by Eq. (4.11) for andisols as well as the histogram of the bias

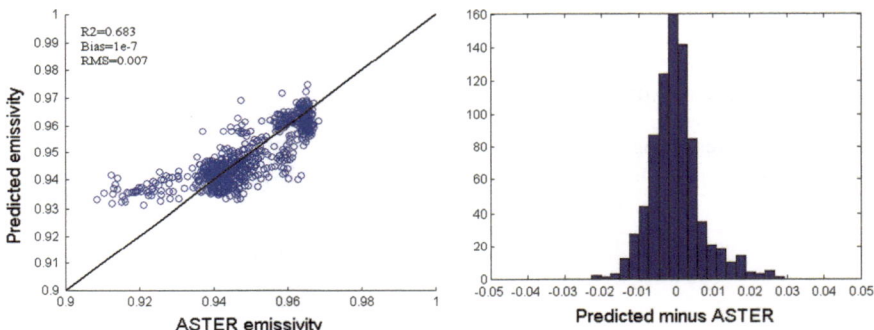

Fig. 4.11 Comparison of ASTER BBE and that predicted by Eq. (4.12) for vertisols as well as the histogram of the bias

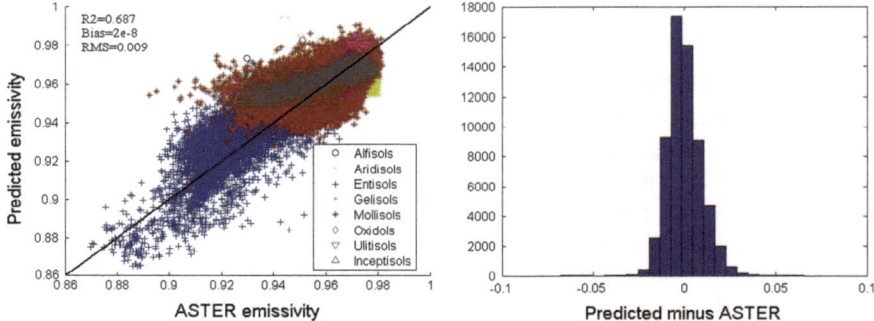

Fig. 4.12 Comparison of ASTER BBE and that predicted by Eq. (4.13) for remaining eight soil orders as well as the histogram of the bias

when generating a global land surface BBE product. It was proposed to specify a transition zone to mitigate the BBE differences between bare soil and vegetated pixels using the NDVI to provide a BBE estimation method for the transition zone. The BBE of the pixels located in the transition zone is the average of their affiliations, which depends on the NDVI of the pixels.

In this study, pixels with an NDVI ranging from 0.1 to 0.2 were labeled as transition zone pixels. If the NDVI is between 0.1 and 0.156, its BBE is the average of that calculated using the equations for bare soil and for the transition zone. If the NDVI lies between 0.156 and 0.2, its BBE is the average of that calculated using the equations for the transition zone and for vegetation.

A sum of 62,879 BBE-albedo pairs was extracted for the transition zone from the same data used to extract the BBE-albedo pairs for bare soil. Similarly to the method used for bare soil, one equation was derived for all soil orders. However, this method did not perform well for all soil orders combined. Therefore, three equations were derived, i.e., one for andisols, one for vertisols, and one for the remaining eight soil orders. Each equation and the coefficients for each variable were significantly below the confidence level of $P < 0.05$. The equations can be expressed as follows:

$$\varepsilon_{BB\_t1} = 1.006 - 0.339a_2 + 0.142a_7 \qquad (4.11)$$

$$\varepsilon_{BB\_t2} = 0.964 + 0.195a_1 + 0.256a_2 - 0.745a_3 + 0.099a_6 - 0.300a_7 \qquad (4.12)$$

$$\varepsilon_{BB\_t3} = 0.954 - 0.782a_1 + 0.345a_2 + 0.776a_3 - 0.111a_4 + 0.056a_5 + 0.080a_6 \\ - 0.131a_7$$

$$(4.13)$$

where ε_{BB_t1}, ε_{BB_t2}, and ε_{BB_t3} are the BBEs for andisols, vertisols, and the remaining eight soil orders, respectively. Figure 4.10 shows a scatter plot of the BBEs predicted using Eq. (4.11) versus the ASTER BBE for andisols and the histogram of the bias. The correlation was relatively low with a value of 0.46, while the bias and RMSE were $-1E^{-7}$ and 0.006, respectively.

The scatter plot of the BBEs predicted using Eq. (4.12) versus the ASTER BBE for vertisols is shown in Fig. 4.11. The correlation was 0.683 while the bias and RMSE were $1E^{-7}$ and 0.007, respectively. The scatter plot of the BBEs predicted using Eq. (4.13) versus the ASTER BBE for the eight soil orders is shown in Fig. 4.12, where the correlation was 0.687. The bias was centered on zero and distributed in a narrow band, but its value was very small and can be neglected, while the RMSE was 0.011. Table 4.4 shows the bias and RMSE for each of the eight soil orders. The absolute bias and RMSE were <0.008 and 0.012, respectively. Equation (4.13) was used to calculate the BBEs of the other soil orders in the soil taxonomy (Table 4.6).

We extracted the BBE-albedo pairs from the test data for nine soil orders. We used these data to evaluate Eqs. (4.11)–(4.13). Figures 4.13, 4.14 and 4.15 show the results of the comparison. The bias and RMSE for Eq. (4.11) were zero and

Table 4.6 Bias and RMS for each of the eight soil orders

Soil order	Bias	RMS
Alfisols	0.008	0.012
Aridisols	−0.002	0.009
Entisols	0.004	0.012
Gelisols	0.000	0.009
Mollisols	0.000	0.009
Oxisols	0.000	0.008
Ultisols	0.000	0.006
Inceptisols	0.005	0.011

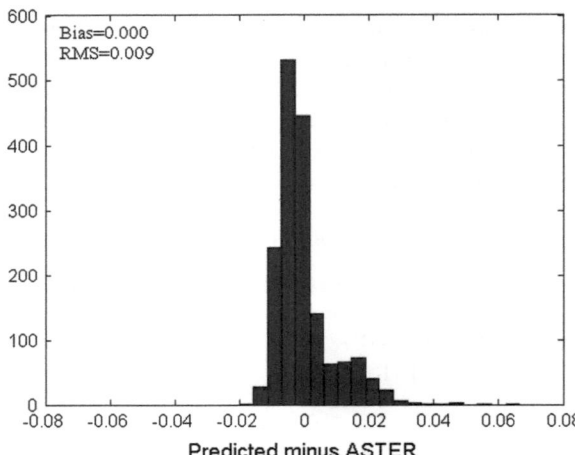

Fig. 4.13 Difference histogram of ASTER BBE and that retrieved by the formula for andisols

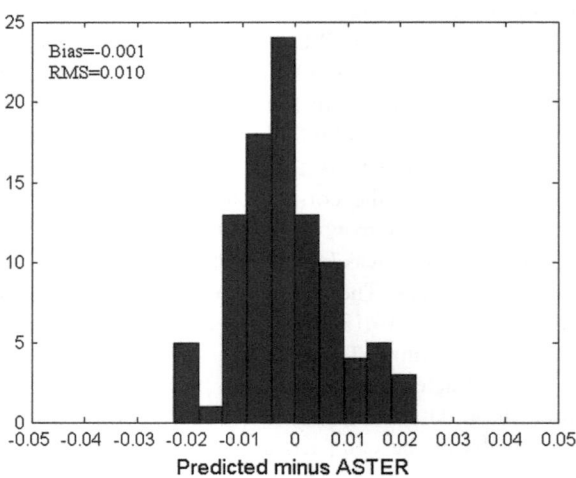

Fig. 4.14 Difference histogram of ASTER BBE and that retrieved by the formula for vertisols

Fig. 4.15 Difference histogram of ASTER BBE and that retrieved by the formula for remaining eight soil orders

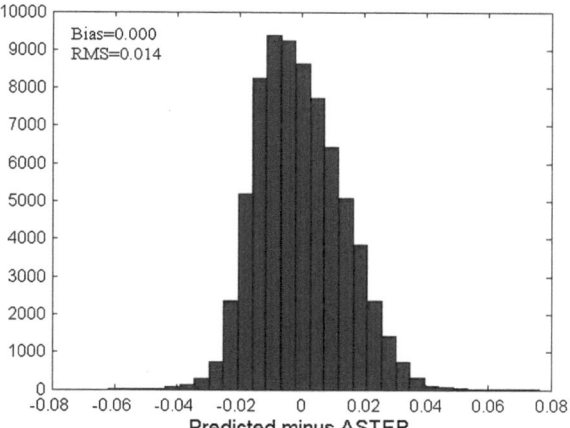

0.014, respectively. With Eq. (4.12), the bias and RMSE were −0.001 and 0.010, while the values for Eq. (4.13) were zero and 0.009, respectively. Overall, the BBE calculated using Eqs. (4.11)–(4.13) agreed well with ASTER BBE and the bias and RMSE were <0.001 and 0.014.

The retrieved BBE values were also compared with that derived from the MODIS emissivity products. According to the validation of different versions of the MODIS Land Surface Temperature and Emissivity (LST&E) products using laboratory-measured sand emissivities, the Version 4.1 emissivity product is the best, whereas the Version 5 emissivity product is the worst (Hulley and Hook 2009a). The estimated BBE values with the BBE values derived from MOD11B1 LST&E products (Versions 4.1 and 5) were compared, which are daily 5- and 6-km products in a sinusoidal projection, retrieved using a day/night algorithm from co-registered day and night image pairs (Wan and Li 1997). To collect as many bare soil pixels as possible, the experimental areas selected for the four soil orders (aridisols, entisols, gelisols, and mollisols) corresponded to large numbers of bare soil pixels as the study areas in this section.

The time intervals for the data were the same as those for the test data in Table 4.4. First, the 8-day MODIS BBE was developed in the following two steps. (1) *Derive the 8-day narrowband emissivity.* The 8-day narrowband emissivity was derived by aggregating the daily emissivity products. (2) *Calculate the 8-day MODIS BBE.* The MODIS BBE was calculated in the 8–13.5 spectral range using Eq. (4.2) from three combined 8-day narrowband emissivities.

Second, the 1 km BBE estimated using the new algorithm was aggregated to 5 and 6 km, respectively.

Finally, the MODIS BBE was compared with the estimated BBE. The results are shown in Figs. 4.16 and 4.17. The MODIS BBE derived from the Version 5 emissivity product was larger than the estimated BBE, which was similar to a previous result (Wang and Liang 2009a). The biases were −0.012 and −0.013 and

Fig. 4.16 Difference
histograms of the MODIS
BBE and of that estimated
using the new algorithm for
bare soil. **a** Version 5;
b Version 4.1

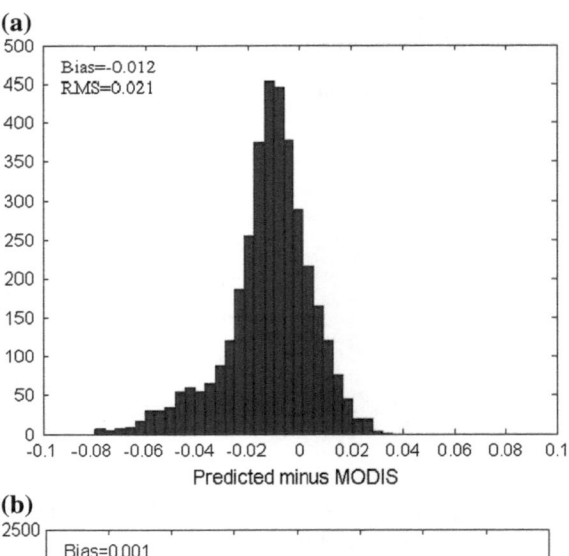

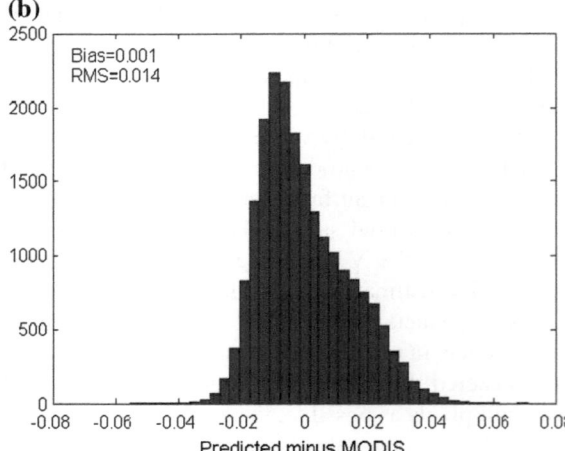

the RMSEs were 0.021 and 0.019 for the bare soil and transition zone, respectively.

The MODIS BBE derived from the Version 4.1 emissivity product was less than the estimated BBE, and the biases were 0.001 and 0.009, while the RMSEs were 0.014 and 0.018 for the bare soil and the transition zone, respectively. The estimated BBE was also compared with the ASTER BBE for these four soil orders. The estimated BBE was in good agreement with the ASTER BBE, i.e., the biases were 0.000 and 0.001 while the RMSEs were 0.014 and 0.014 for the bare soil and the transition zone, respectively.

Therefore, two conclusions can be drawn based on a comparison of the results. The first is that the BBE values derived from different versions of the MODIS

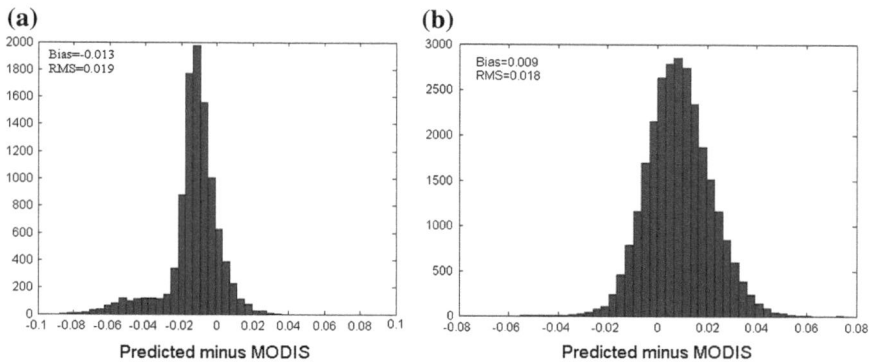

Fig. 4.17 Difference histograms of the MODIS BBE and of that estimated using the new algorithm for the transition zone. **a** Version 5; **b** Version 4.1

emissivity product were inconsistent. The second is that the accuracy of the estimated BBE is higher than those derived from the MODIS emissivity products.

The scheme for retrieving global land surface BBE for vegetated area is shown in Fig. 4.18. The BBEs are calculated from five simultaneous TIR channel LSEs

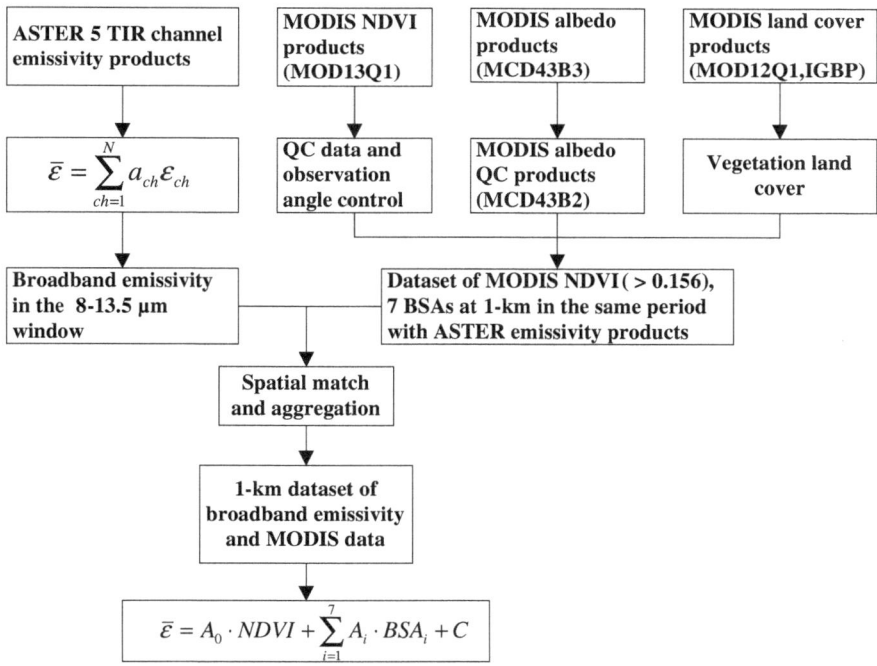

Fig. 4.18 Scheme for deriving the relationship from MODIS NDVI, seven BSAs to BBE (Ren et al. 2013). Copyright © 2013 Reproduced with permission of IEEE

found in the ASTER product AST05 using Eq. (4.7). It was necessary to aggregate the 90 m ASTER data and the 250 m MODIS NDVI to that scale by averaging 11 × 11 pixels of ASTER data (4 × 4 pixels of NDVI). The averaging process was constrained by the degree of heterogeneity, which was defined such that the standard deviation of BBEs in the 11 × 11 window (NDVI in 4 × 4) would not be greater than 0.015 (0.03). Finally, a linear relationship was assumed to link BBEs to the MODIS NDVI and the seven channel BSAs (CH1: 620–670 nm, CH2: 841–876 nm, CH3: 459–479 nm, CH4: 545–564 nm, CH5: 1,230–1,250 nm; CH6: 1,628–1,652 nm; and CH7: 2,105–2,155 nm), expressed as:

$$\bar{\varepsilon} = A_0 \cdot \text{NDVI} + \sum_{i=1}^{7} A_i \cdot \text{BSA}_i + C \qquad (4.14)$$

where A_0 and A_i are coefficients, and C is a constant.

The ASTER LSE products and their corresponding MODIS NDVI, spectral albedos, and albedo QC (MCD43B2) products were collected along with land cover products at 40 sites around the world (red points in Fig. 4.19) for 2000 to 2010, and 1,012 images of those products under cloud-free sky were obtained. A total of 43,613 vegetation samples (shrubland, savanna, grassland, cropland, and forest) were chosen from those data at a 1 km scale.

Table 4.7 shows the number of samples and their percentages, indicating that each sample was large enough to ensure a representative result. The coefficients A_0, A_i, and C in Eq. (4.14) were determined by stepwise regression and are listed in Table 4.8. All the variables except BSA_2 were statistically significant (t test with $p < 0.01$). The RMSE of the residual error from the linear regression was 0.01, but the bias is not listed because it was very close to zero.

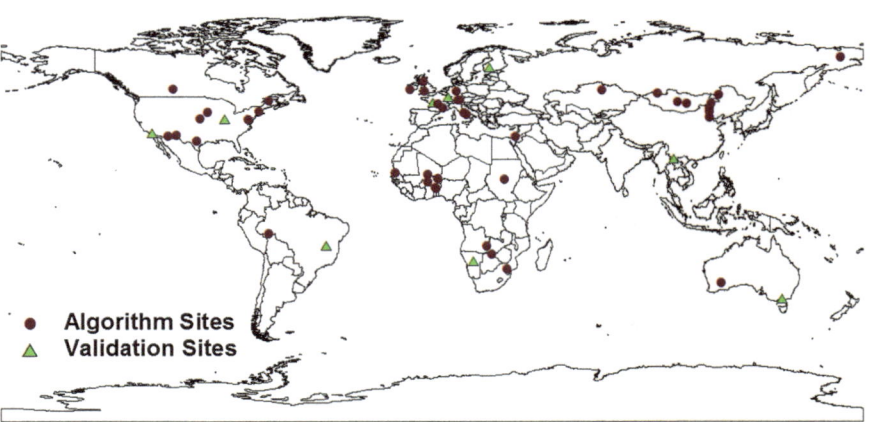

Fig. 4.19 Distribution of sites for algorithm development and validation. *Red* points represent sites where data were used to develop algorithms while *triangle* points represent sites where data were used to carry out validation of the algorithms (Ren et al. 2013). Copyright © 2013 Reproduced with permission of IEEE

Table 4.7 Different types of vegetation samples (total: 43613)

Land cover types	Number of samples	Percentage (%)
Shrubland	11,036	25.3
Savannas	9,397	21.5
Grassland	9,156	21.0
Cropland	8,453	19.4
Forest	5,571	12.8

Table 4.8 Regression coefficients for BBE in the 8–13.5 μm spectral range (with t test, $p < 0.01$)

A_0	A_1	A_2	A_3	A_4	A_5	A_6	A_7	C	RMSE	R^2
0.033	−0.066	0.0	−0.291	0.551	−0.324	0.411	−0.373	0.977	0.011	0.55

It was found that most of the original and fitted BBEs were between 0.92 and 0.985, which was within the range of the measured emissivity for vegetated surfaces (Van de Griend et al. 1991), that is, from 0.94 for sorghum with 10 % coverage of bare soil to 0.986 for complete shrub cover. The histograms of the residual errors between the fitted and the original BBEs presented in Fig. 4.20 show that approximately 66 % of the total pixel residual errors lay between −0.01 and 0.01, and 83.3 % of the error between −0.015 and 0.015. The results indicate that the algorithm is able to retrieve BBEs with a relative error less than 0.015 for most cases. This accuracy is close to the requirement for current LSE products reported in a previous study (Gustafson et al. 2006).

For simplicity and convenience, here provides only the regression results for all vegetated pixels. However, it is necessary to test the coefficients for each type of vegetation cover. Table 4.9 shows the statistical results of the BBE residual errors for each vegetation cover types, with their average NDVI and standard deviation.

Fig. 4.20 Histograms of BBE residual errors (Ren et al. 2013). Copyright © 2013 Reproduced with permission of IEEE

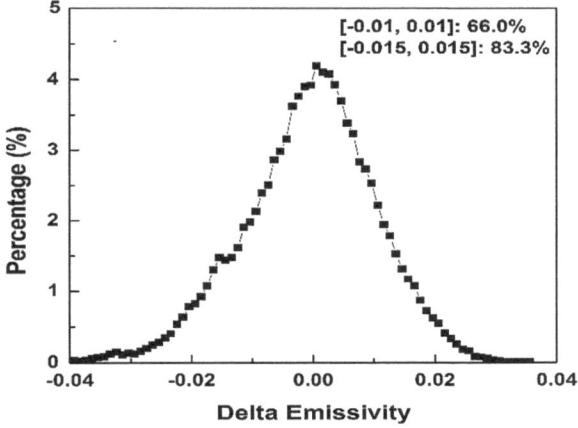

Table 4.9 Comparison of fitted and original BBEs for each vegetation cover

Land cover type	NDVI (std.)	RMSE	[−0.01, 0.01] (%)	[−0.015, 0.015] (%)
Shrubland	0.213 (0.066)	0.010	65.1	84.6
Grassland	0.267 (0.143)	0.009	73.6	88.4
Forest	0.804 (0.121)	0.007	84.3	96.5
Savannas	0.418 (0.144)	0.012	63.1	78.3
Cropland	0.454 (0.245)	0.010	62.4	82.3
Total	0.345 (0.209)	0.011	66.0	83.3

It indicates that the RMSEs of the residual BBEs for shrubland, grassland, forest, and cropland were not greater than 0.01. The cumulative percentages of the residual errors within [−0.01, 0.01] and [−0.015, 0.015] for these four covers were, respectively, similar or very close to those of the overall case. Moreover, forest pixels had the minimum RMSEs (0.007) and the largest percentages within [−0.01, 0.01] and [−0.015, 0.015], probably because little variation occurs in their coverage over a year, especially for evergreen forests.

Comparison with ASTER products over different places and at different times helps to test the feasibility of the proposed algorithm for vegetation area. Consequently, 126 clear-sky images of the ASTER LSE products, and the MODIS NDVI, albedo, and albedo QC products from 2000 to 2010 were collected over an additional nine sites (the triangles in Fig. 4.19). According to the schema in Fig. 4.18 and the coefficients in Table 4.8, BBEs were obtained at 1 km spatial resolution, respectively, from MODIS data using the MDBA, and from ASTER products by converting their channel LSEs to broadband using the coefficients in Table 4.11. The 1 km resolution ASTER BBEs were aggregated from the average of approximately 11 × 11 90 m-resolution BBE pixels with a degree of heterogeneity (defined above) less than 0.015.

Table 4.10 summarizes the results of this comparison. It shows that the RMSEs (column 4) of all sites except Omatako Ranch (grassland) and Pandeiros (savanna) were less than 0.01. Omatako Ranch is in southern Africa with sparse coverage so that the background soil probably contributes to its BBE. In the method proposed

Table 4.10 Validation results with ASTER emissivity products at nine validation sites

Site name	Main cover[a]	Latitude/longitude (°)	RMSE
Omatako Ranch	Grassland	−21.3/17.4	0.010
Sky Oaks	Shrubland	33.3/−116.6	0.008
Lusignan	Cropland	46.3/0.5	0.008
Morgan Forest	Deci BL Forest	39.3/−86.4	0.007
Hesse Forest	Deci BL Forest	48.7/7.1	0.008
Wallaby Creek	Evgn BL Forest	−37.6/146.3	0.008
Xishuang-banna	Evgn BL Forest	22.0/101.1	0.007
Hyytiala	Evgn NL Forest	61.9/24.3	0.007
Pandeiros	Savannas	−14.9/−44.0	0.011

[a] *Deci* Deciduous, *BL* Broadleaf, *NL* Needleleaf, *Evgn* Evergreen

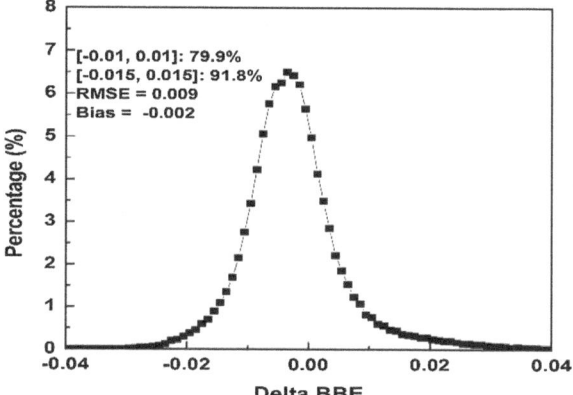

Fig. 4.21 BBE difference histograms at all nine validation sites (Ren et al. 2013). Copyright © 2013 Reproduced with permission of IEEE

here, this might lead to larger relative errors. In addition, as shown in Table 4.10, because the algorithm is not very robust for savanna, it is not surprising that the Pandeiros site is associated with the highest error, 0.011. Figure 4.21 shows the BBE difference histogram at all nine validation sites. The RMSE was approximately 0.009, and the bias was approximately −0.002. The cumulative percentages of their differences were approximately 79.9 and 91.8 % within [−0.01, 0.01] and [−0.015, 0.015], respectively. These comparisons indicate that the regressed coefficients in the proposed model can be used to estimate BBE within acceptable errors.

4.2.3 *Estimating Broadband Emissivity from AVHRR VNIR Data*

The algorithm for retrieving global land surface BBE from AVHRR VNIR data was similar to that for retrieving global land surface BBE from MODIS albedos (Cheng and Liang 2013b). The flowchart of the algorithm is presented in Fig. 4.22. Both AVHRR VNIR data and BBE data derived from MODIS albedos from 2000 were acquired. AVHRR VNIR data and BBE were spatially and temporally matched. The 8-day 1 km BBE with a sinusoidal grid projection was reprojected to 0.05° CMG, and then daily reflectance data of AVH09 channels 1 and 2 were combined at an 8-day step size. Soil taxonomy was also reprojected to 0.05° CMG. High-quality clear-sky reflectance-BBE pairs were extracted from spatially and temporally matched data for each land surface type, which was identified by NDVI calculation from the combined 8-day AVHRR VNIR reflectance. Similarly to the algorithm used to derive BBE from MODIS albedos, land surface was divided into three types according to NDVI. The NDVI threshold values used were based on previous studies (Momeni and Saradjian 2007; Sobrino et al. 2008). A bare soil

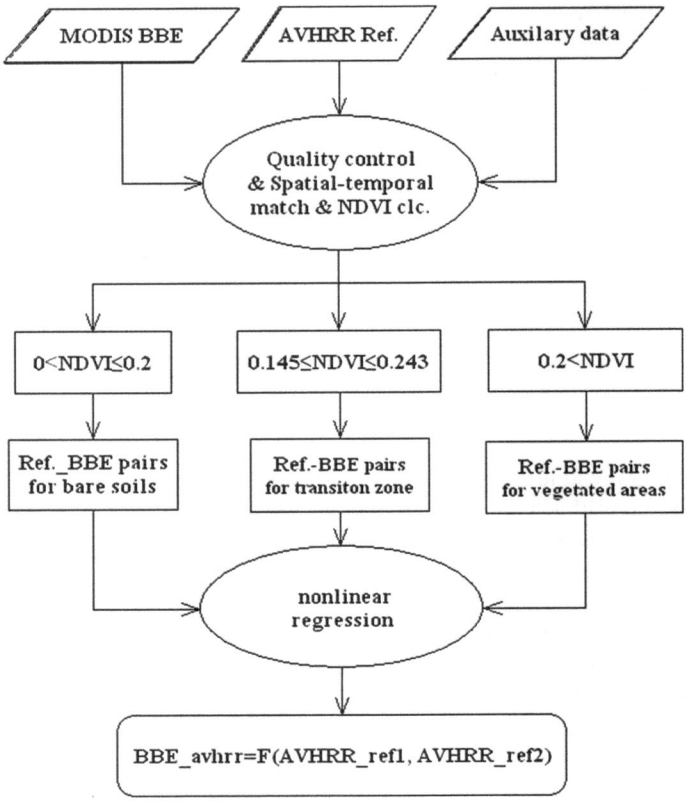

Fig. 4.22 Flowchart of the relationship established between MODIS BBE and AVHRR reflectance (Cheng and Liang 2013b)

pixel was identified as $0 < \text{NDVI} \leq 0.2$, a transition zone pixel as $0.145 \leq \text{NDVI} \leq 0.243$, and a vegetated area pixel as $\text{NDVI} \geq 0.2$. Finally, nonlinear equations for bare soil, vegetated areas, and transition zones were established using nonlinear regression with extracted reflectance-BBE pairs.

The AVHRR data used in this study were obtained from the AVH09 Surface Reflectance Product (Version 3) generated by the Land Long-Term Data Record (LTDR) project, the goal of which is to produce a consistent long-term data set from AVHRR and MODIS for use in global change and climate studies (Pedelty et al. 2007). The temporal resolution of AVH09 is 1 day. Processed AVHRR (AVH09) observations were grouped into separate scientific data sets (SDSs) in a single hierarchical data format file, covering a 0.05° spatial resolution in a latitude/longitude climate modeling grid (CMG). These SDSs included surface reflectance for channels 1 (0.5–0.7 μm), 2 (0.7–1.0 μm), and 3 (3.55–3.93 μm) and TOA brightness temperature channels 3 (3.55–3.93 μm), 4 (10.3–11.3 μm), and 5 (11.5–12.5 μm). Reflectance and brightness temperature were atmospherically

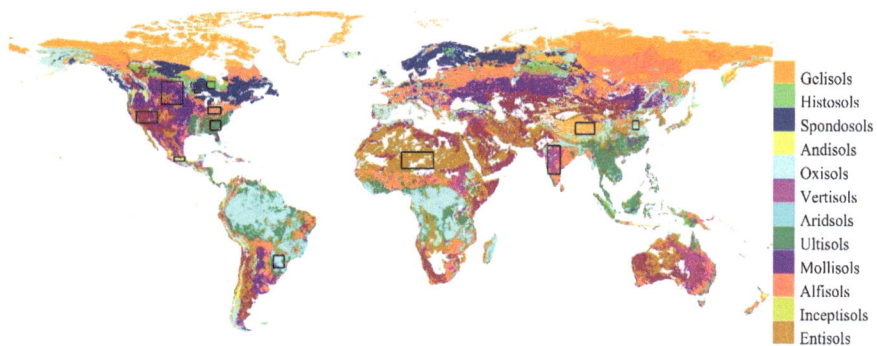

Fig. 4.23 USDA NRCS soil taxonomy and selected study areas. Other classes, such as rocky land, shifting sand, and ice or glacier, are not included in the map (Cheng and Liang 2013b). Copyright © 2013 Reprinted by permission of Taylor & Francis Ltd

corrected, along with Rayleigh scattering, ozone, water vapor, and aerosols. Soil taxonomy and the MODIS land cover product (MOD12C1) were the required auxiliary data. The former was used to identify different soil orders, whereas the latter was used to identify different vegetation types. Selected study areas (black rectangles in Fig. 4.23) for bare soil and transition zones were the same. Relatively large homogeneous areas were selected for each soil order to obtain more representative samples for bare soil. The geographical locations of the data used are presented in Table 4.11. Reflectance-BBE pairs extracted from the data for days 49, 65, 81, 97, 105, 145, 241, and 321 in 2000 were used to establish nonlinear relationship between BBE and VNIR reflectance. Reflectance-BBE pairs extracted from the data for day 209 in 2000 were used to test the established nonlinear relationship. Based on MOD12C1 data, 49 sites were selected from MODIS land subsets.

Table 4.11 Geographic locations from which data used to develop the BBE retrieval algorithm for both bare soils and transition zones were derived

Soil order	Alfisols	Andisols	Aridisols	Entisols	Gelisols	Inceptisols
Geographic location	39° to 42°	18.5° to 20.5°	35° to 40°	16° to 23°	31° to 36°	33° to 37°
(latitude, longitude)	−86° to −80°	−102° to −97°	−120° to −110°	5° to 20°	87° to 96°	114° to 117°
Soil order	Mollisols	Oxisols	Ultisols	Vertisols	Spodosols	Histisols
Geographic location	43° to 3°	−27.5° to −22°	32° to 36°	14° to 26°	48° to 51°	50° to 53°
(latitude, longitude)	−108° to −98°	−55° to −50°	−85° to −80°	74° to 80°	−92.5° to 89.5°	−86.6° to −82.8°

The distribution of the selected sites is shown in Fig. 4.19. For each site, all data in 2000 were extracted from the $1° \times 1°$ square region centered on this site. Among these sites, 40 were used to establish a nonlinear relationship between BBE and VNIR reflectance, and nine were used to test the established relationship. Land cover for both development and test sites included forest, cropland, grassland, savanna, and shrub.

For bare soil, 501,815 samples were obtained from the study areas. One equation was derived for all samples. However, this equation could not fully characterize the relationship between B9BE and reflectance. Therefore, two equations were derived: one for individual vertisols and another for the remaining 11 soil orders. Each equation and the coefficient for each variable were significant at the 0.05 confidence level. The equations can be expressed as follows:

$$\varepsilon_{\mathrm{BB}\_s1} = 0.957 + 0.179R_2 - 0.822R_2^2, \tag{4.15}$$

$$\varepsilon_{\mathrm{BB}\_s2} = 0.988 - 0.734R_1 + 0.477R_2 + 1.069R_1^2 - 0.783R_2^2, \tag{4.16}$$

where $\varepsilon_{\mathrm{BB}\_s1}$ and $\varepsilon_{\mathrm{BB}\_s2}$ are BBEs for vertisols and the remaining 11 soil orders, respectively; and R_1 and R_2 are reflectance of AVHRR channels 1 and 2, respectively. The BBE scatter plot predicted using Eq. (4.15) versus BBE derived from MODIS albedos for vertisols and the histogram of bias are shown in Fig. 4.24. The predicted BBE was less than 0.97. Most points were distributed around the 1:1 line. However, the points were highly dispersed when the BBE was less than 0.96. Therefore, the overall fit was not very good, with a correlation of only 0.529, but the bias (as small as $4\mathrm{E} - 5$) was centered on zero and distributed in a narrow band, and the RMSE was 0.004. The BBE scatter plot predicted by Eq. (4.16) versus the BBE derived from MODIS albedos for 11 soil orders and the histogram of bias are shown in Fig. 4.25. Most points were distributed around the 1:1 line. The correlation was 0.822. The bias (-0.001) was also centered on zero and distributed in a narrow band, and the RMSE was 0.012.

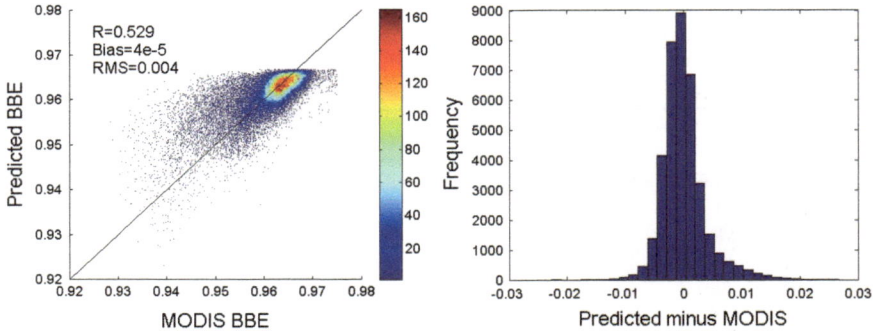

Fig. 4.24 Scatter plot and difference histogram of BBE derived from MODIS albedos and BBE calculated using Eq. (4.15) (Cheng and Liang 2013b). Copyright © 2013 Reprinted by permission of Taylor & Francis Ltd

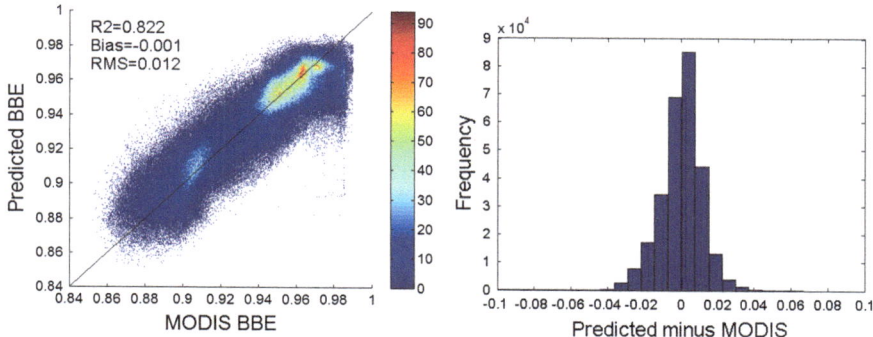

Fig. 4.25 Scatter plot and difference histogram of BBE derived from MODIS albedos and BBE calculated using Eq. (4.16). From Cheng and Liang (2013b). Copyright © 2013 Reprinted by permission of Taylor & Francis Ltd

A total of 92,691 samples were obtained to establish a nonlinear relationship between BBE and reflectance for vegetated areas. The BBE for vegetated areas was formulated as follows:

$$\varepsilon_{BB\_v} = 0.962 + 0.125R_1 + 0.043R_2 + 0.457R_1^2 - 1.323R_1R_2 + 0.107R_2^2, \quad (4.17)$$

where ε_{BB_v} is the BBE for vegetated areas. The scatter plot of BBE predicted using Eq. (4.17) versus BBE derived from MODIS albedos and the bias histogram are shown in Fig. 4.26. BBE was clearly underestimated when it was greater than 0.98. The correlation was 0.588. The bias (as small as 2E − 5) was centered on zero and distributed in a narrow band, and the RMSE was 0.007.

An equation was derived for all soil orders by applying the same method used for bare soil. However, this equation did not work well for all soil orders, and therefore, two equations were derived: one for vertisols and another for the

Fig. 4.26 Scatter plot and difference histogram of BBE derived from MODIS albedos and BBE calculated using Eq. (4.17) (Cheng and Liang 2013b). Copyright © 2013 Reprinted by permission of Taylor & Francis Ltd

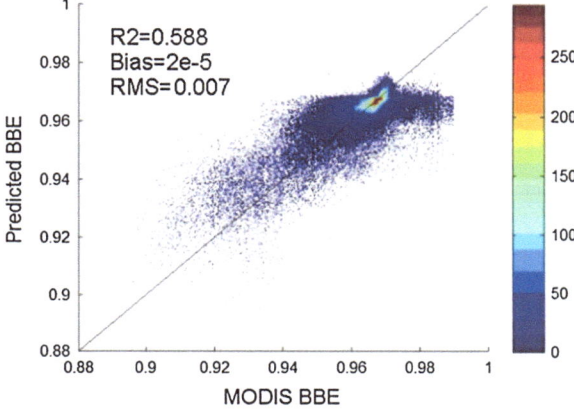

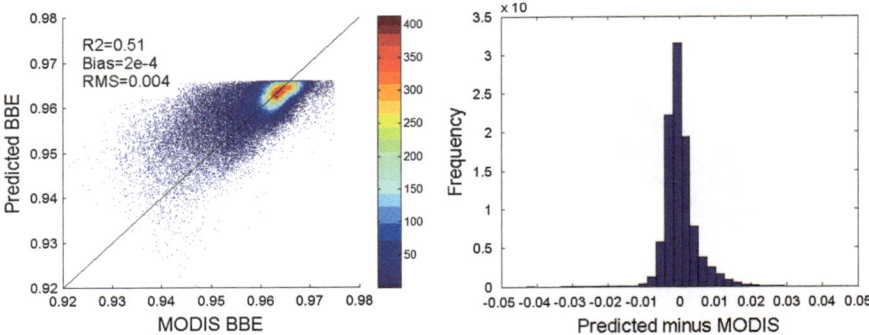

Fig. 4.27 Scatter plot and difference histogram of BBE derived from MODIS albedos and BBE calculated using Eq. (4.18) (Cheng and Liang 2013b). Copyright © 2013 Reprinted by permission of Taylor & Francis Ltd

remaining 11 soil orders. Each equation and each variable coefficients were significant at the 0.05 confidence level. The equations can be expressed as follows:

$$\varepsilon_{BB\_t1} = 0.955 + 0.185R_2 - 0.78R_2^2 \tag{4.18}$$

$$\varepsilon_{BB\_t2} = 0.972 - 0.374R_1 + 0.297R_2 + 0.362R_1^2 - 0.52R_2^2 \tag{4.19}$$

where ε_{BB_t1} and ε_{BB_t2} are BBEs for vertisols and the remaining 11 soil orders, respectively. The BBE scatter plot predicted using Eq. (4.18) versus BBE derived from MODIS albedos for vertisols and the bias histogram are shown in Fig. 4.27. The predicted BBE was less than 0.97. The points were dispersed when the BBE was less than 0.96. The correlation coefficient was 0.51. The bias (as small as $2E - 5$) was centered on zero and distributed in a narrow band. The RMSE was

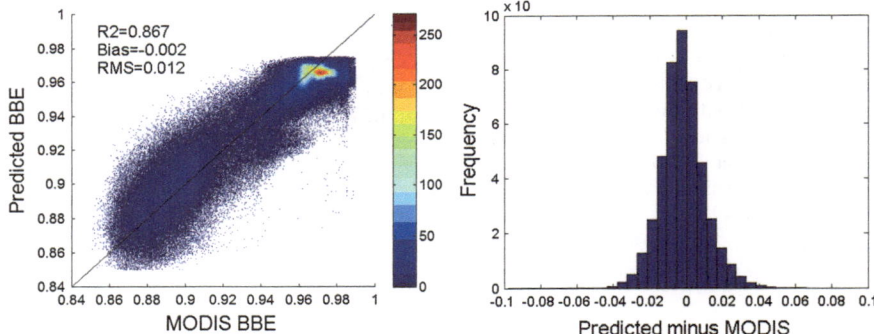

Fig. 4.28 Scatter plot and difference histogram of BBE derived from MODIS albedos and BBE calculated using Eq. (4.19) (Cheng and Liang 2013b). Copyright © 2013 Reprinted by permission of Taylor & Francis Ltd

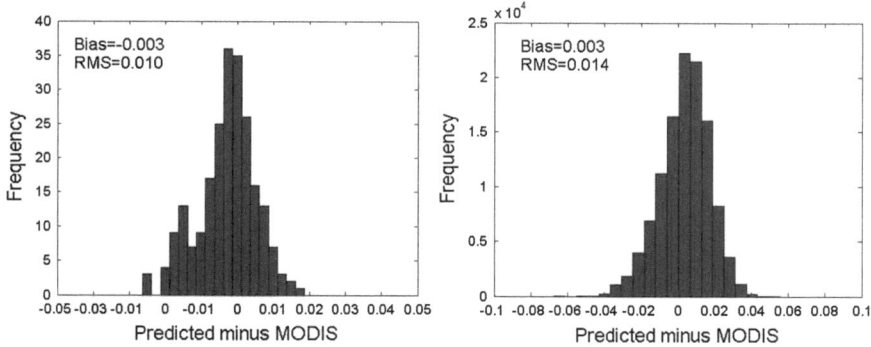

Fig. 4.29 Difference histograms of BBE derived from MODIS data and BBE calculated by Eqs. (4.15) (*left*) and (4.16) (*right*) (Cheng and Liang 2013b). Copyright © 2013 Reprinted by permission of Taylor & Francis Ltd

0.004. The BBE scatter plot predicted by Eq. (4.19) versus the BBE derived from MODIS albedos for the 11 soil orders and the bias histogram are shown in Fig. 4.28. Most points were distributed around the 1:1 line. The correlation coefficient was 0.867, bias was −0.002, and RMSE was 0.012.

The established equations for bare soil, transition zones, and vegetated areas were examined based on extracted test data. Test results for bare soil and transition zones are presented in Figs. 4.29 and 4.30. For bare soil, absolute bias and RMSE were less than 0.003 and 0.014, respectively. For transition zones, absolute bias and RMSE were less than 0.002 and 0.011, respectively. Test results for vegetated areas are shown in Fig. 4.31, with a bias and RMSE of −0.002 and 0.005, respectively. The retrieved BBE values agreed with the BBE values derived from MODIS data, thus indicating that the established equations can reflect the relationship between the BBE and VNIR reflectance.

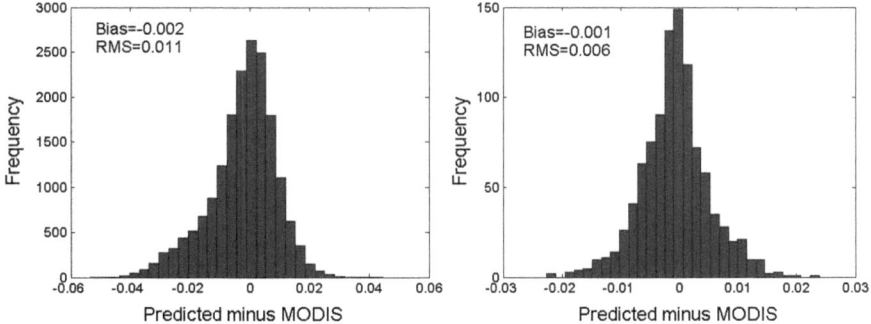

Fig. 4.30 Difference histograms of BBE derived from MODIS data and BBE calculated by Eqs. (4.18) (*left*) and (4.19) (*right*) (Cheng and Liang 2013b). Copyright © 2013 Reprinted by permission of Taylor & Francis Ltd

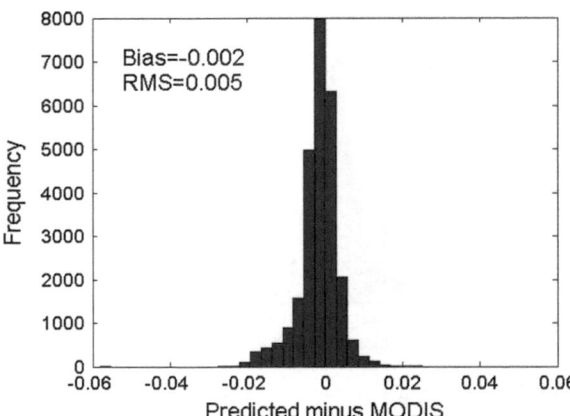

The algorithm was also tested by comparing the retrieved BBE values with those retrieved by the NDVI threshold method. Based on Botswana's radiometer-measured data, which correspond with the AVHRR channel, Van de Griend, and Owe (1993) found that LSE has a good logarithmic relationship with NDVI. Through further analysis of the relationship between LSE and NDVI, Olioso (1995) noted that this dependence is closely related to soil emissivity, component effective emissivity of leaves, canopy structure, optical features of leaves, solar position, and proportion of soil penetrated by sunlight, although it is insensitive to observational geometric conditions. Valor and Caselles (1996) used a vegetation cover method to calculate LSE and applied this method to more complex mixed pixels to obtain satisfactory results. Sobrino et al. (2001, 2008) proposed an NDVI threshold method for AVHRR data, the mathematical expressions of which are presented as follows:

$$\begin{cases} \varepsilon_4 = 0.979 - 0.057R_1 & \text{NDVI} < 0.2 \\ \varepsilon_4 = 0.968 + 0.021P_v & 0.2 \leq \text{NDVI} \leq 0.5, \\ \varepsilon_4 = 0.99 & \text{NDVI} > 0.5 \end{cases} \qquad (4.20)$$

$$\begin{cases} \varepsilon_5 = 0.982 - 0.028R_1 & \text{NDVI} < 0.2 \\ \varepsilon_5 = 0.974 + 0.015P_v & 0.2 \leq \text{NDVI} \leq 0.5, \\ \varepsilon_5 = 0.99 & \text{NDVI} > 0.5 \end{cases} \qquad (4.21)$$

where ε_4 and ε_5 are emissivities for the AVHRR channels 4 and 5; $P_v = \left(\frac{\text{NDVI}-\text{NDVI}_s}{\text{NDVI}_v-\text{NDVI}_s}\right)^2$ is the fractional vegetation cover; NDVI_s is the NDVI of soil; and NDVI_v is the NDVI of vegetation. A pixel represents soil when $\text{NDVI} < \text{NDVI}_s$. The pixel is a mixture of vegetation and soil (i.e., has partial vegetation cover) when $\text{NDVI}_s \leq \text{NDVI} \leq \text{NDVI}_v$. The pixel represents vegetation when $\text{NDVI} > \text{NDVI}_v$. A linear function was derived from the ASTER spectral library to convert AVHRR narrowband emissivity to broadband emissivity at

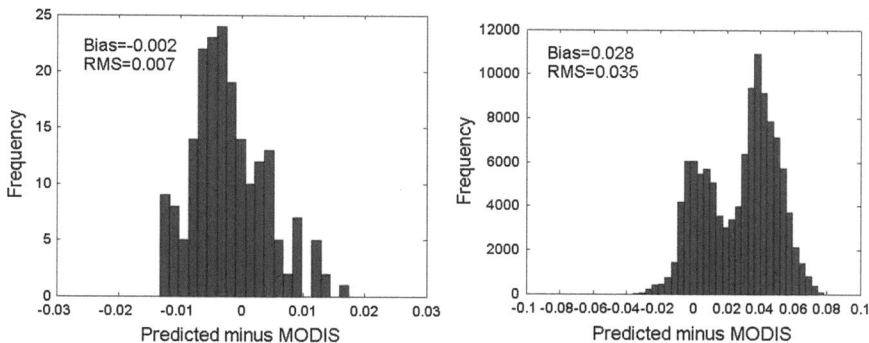

Fig. 4.32 Difference histograms of the BBE values derived from MODIS albedos and those derived from the NDVI threshold method for vertisols (*left*) and for the remaining 11 soil orders (*right*) (Cheng and Liang 2013b). Copyright © 2013 Reprinted by permission of Taylor & Francis Ltd

8–13.5 μm. This function is presented in Eq. (4.8). R-squared was 0.60 and the RMSE was 0.018:

$$\varepsilon_{BB} = 0.305 + 0.674\varepsilon_4 \tag{4.22}$$

Channel 4 emissivity was extracted from the test data and converted into BBE using Eq. (4.22). The derived BBE values were compared with those retrieved from MODIS data. Comparison results are shown in Figs. 4.32 and 4.33. For vertisols, the BBE values retrieved using the NDVI threshold method were closer to those derived from MODIS data, with bias and RMSE of −0.002 and 0.007, respectively. However, high bias and RMSE were observed for the remaining 11 soil orders, with values as high as 0.028 and 0.035, respectively. BBEs were also extracted for partially and fully vegetated areas and compared with those derived

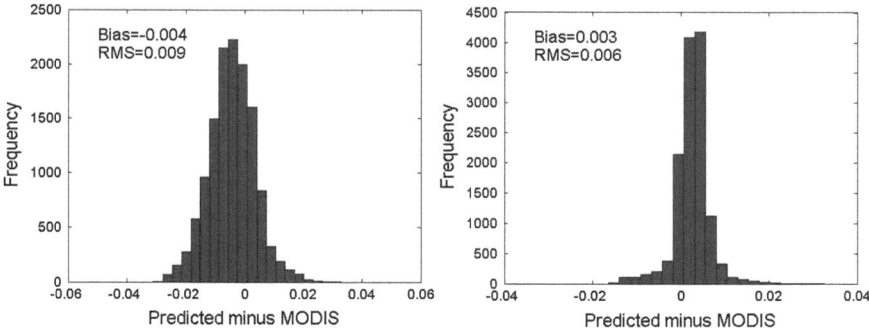

Fig. 4.33 Difference histograms of the BBE values derived from MODIS albedos and those derived using the NDVI threshold method for partially (*left*) and fully vegetated areas (*right*) (Cheng and Liang 2013b). Copyright © 2013 Reprinted by permission of Taylor & Francis Ltd

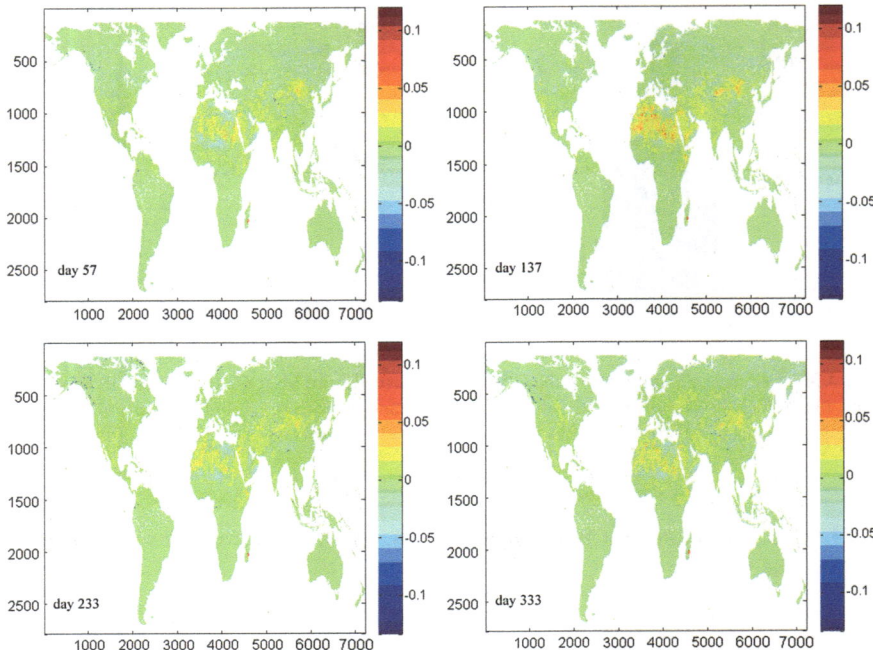

Fig. 4.34 Difference map of global land surface BBE derived from AVHRR VNIR data and MODIS albedos for days 57, 137, 233, and 333 in 2010 (Cheng and Liang 2013b). Copyright © 2013 Reprinted by permission of Taylor & Francis Ltd

from MODIS data. Bias and RMSE were less than 0.004 and 0.009, respectively. A comparison of Figs. 4.29–4.32 reveals that the BBE values retrieved by the nonlinear equations were better than those derived from the NDIV threshold method, especially for bare soil. The global land surface BBE values were retrieved by nonlinear functions developed from AVHRR VNIR data for days 57, 137, 233, and 333 in 2000, the values of which were compared with those derived from MODIS albedos. Difference maps are presented in Fig. 4.34. The differences in the Sahara desert and in desert and semi-arid areas in northwestern China were larger than in other areas. Two AVHRR VNIR channels were obtained to calculate BBE, whereas seven VNIR channels were obtained for MODIS. The information provided by the two sensors was asymmetrical; thus, the difference in retrieved BBEs was unavoidable, albeit very small. Mean values were −0.0008, −0.0006, −0.0005, and 0.0006. RMSEs were 0.008, 0.008, 0.007, and 0.009.

4.3 Product Characteristics, Quality Control, and Validation

4.3.1 Product Characteristics

The developed algorithms were used to generate global land surface 8-day 1 and 5 km BBE product. The data required to produce the global 8-day 5 km land surface BBEs were the AVH09 Surface Reflectance Product (Version 3) and soil taxonomy, and the data required to produce the global 8-day 1 km land surface BBEs were the MODIS reflectance data (MOD09A1), MODIS albedo products (MCD43B3 and MCD43B2), and soil taxonomy. The AVHRR and MODIS data were all processed by another group in our team to ensure good data quality. Processing included cloud clearing, spatiotemporal filtering, and gap filling.

Overall, the land surface was divided into six types: water, snow/ice, bare soil, soil transition zone, vegetation transition zone, and vegetated area for generation of the global land surface BBE product. The emissivity spectrum of water can be simulated using the Fresnel equation given its refractive index. Figure 4.35 shows the simulated pure water emissivity spectra together with the emissivity spectra in the ASTER spectral library and the MODIS UCSB spectral library. The measured emissivity spectra were distributed in a narrow band and agree well with the simulated emissivity spectrum. The BBE of water in the ASTER spectral library was 0.984 for all three samples, while the value in the MODIS UCSB spectral library was 0.985 for all five samples. The BBE of water was specified as 0.985 during production of the global land surface BBE product.

For snow and ice, the radiative transfer model and Fresnel equation were used, respectively, to simulate their emissivity spectra (Cheng et al. 2010b). However, it was impractical to obtain the model inputs (snow effective radius and the

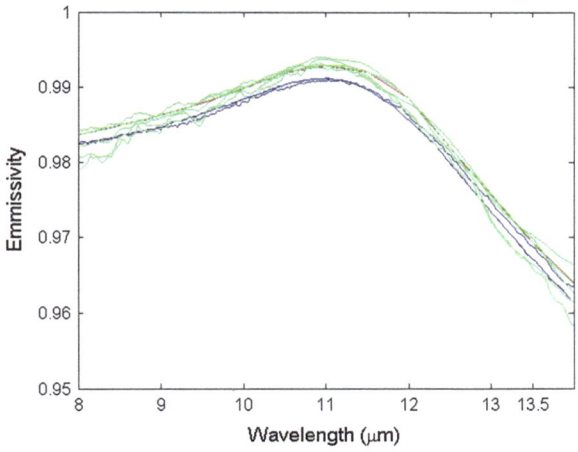

Fig. 4.35 Comparison of simulated pure water emissivity (*red*), water emissivity in the ASTER spectral library (*blue*) and water emissivity in the MODIS UCSB emissivity spectra (*green*)

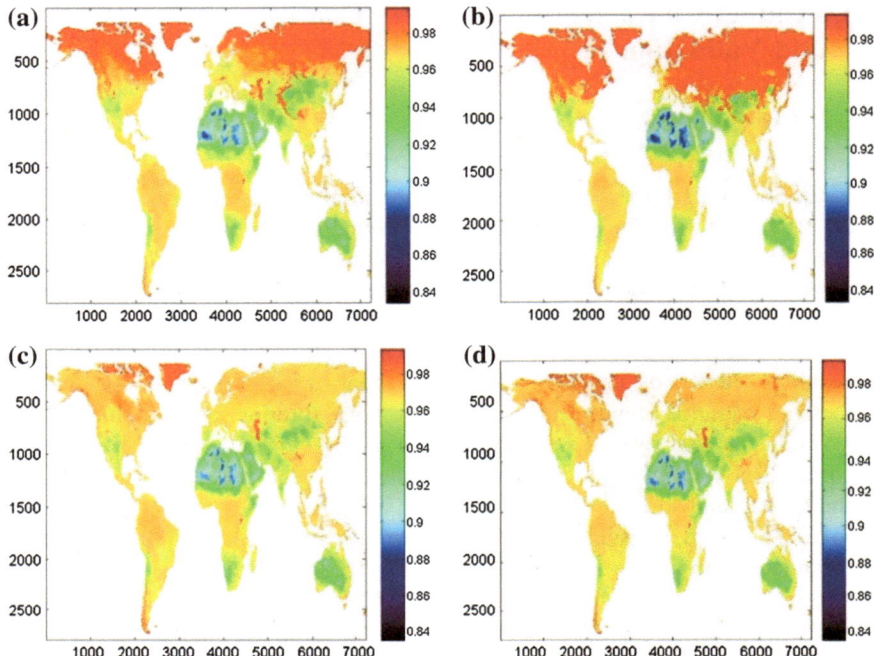

Fig. 4.36 Comparison between the NAALSED BBE and the UWIREMIS BBE for the winter season. **a** UWIREMIS BBE; **b** NAALSED BBE; **c** difference between the NAALSED BBE and the UWIREMIS BBE; **d** histogram of the difference

refractive indices of snow and ice) on a global scale. Moreover, the emissivity spectra of snow/ice were angle dependent and this dependency cannot be well simulated by radiative transfer models (Cheng et al. 2010b; Hori et al. 2006). The BBE values of snow/ice were simulated using the emissivity spectra in the ASTER spectral library and the MODIS UCSB spectral library, and 0.985 was set to their BBE. The error was less than 0.005 when the viewing angle was less than 45°. For other remaining surface types, the developed algorithms were used to retrieve their BBE values.

The 8-day 1-km sinusoidal projection GLASS BBE in 2003 was projected to the 0.05° CMG grid. The monthly BBE for January, April, July, and October was calculated by averaging the GLASS BBE values for these months, as shown in Fig. 4.36. In general, the BBE values were very low over arid and semi-arid areas, for example, the Sahara Desert, the northwestern China, and the western United States. The vegetated areas had relatively high BBE. Most areas of Greenland were covered by snow in April and October, but most snow had melted by July. The snow cover variation for the four seasons was reflected from the GLASS BBE variations. High latitude areas were covered by snow in January, but the snow

Table 4.12 Quality control flags of the GLASS LAI product

Bits (right to left)	Value (defuat:0)	Description
0	0	No emissivity for current pixel
	1	Emissivity was calculated
1	0	Current pixel in MOD09A1 is not preprocced
	1	Current pixel in MOD09A1 is preprocced
2	0	Current pixel in MCD43B3 is not preprocced
	1	Current pixel in MCD43B3 is preprocced
3	0	Variance of NDVI values for the four pixels in MOD09A1 at 500 m is no more than 0.1
	1	Variance of NDVI values for the four pixels in MOD09A1 at 500 m is more than 0.1
4	0	Retrieved emissivity
	1	Emissivity = 0.85 if the retrieved 8–13.5 μm emissivity is less than 0.85
5	0	Retrieved emissivity
	1	Emissivity = 0.99 if the retrieved 8–13.5 μm emissivity is greater than 1.0
6–7	Unused	

began to melt with the passage of time. There was almost no snow cover in July except in Greenland. By October, snow fall was present at high latitudes.

4.3.2 Quality Control

Every scene of the GLASS BBE products has been subjected to a strict quality control (QC) procedure. QC has been carried out both automatically and with human involvement. The automatic quality control refers mainly to the generation of the QC flag, by which the uncertainty of the GLASS BBE products was estimated. The quality check with human involvement is a computer-aided visual inspection of the spatial and temporal patterns of the product before its final release.

The BBE values were stored as 16-bit signed integer data type. Valid product values range from 0 to 10,000, and the scale factor that converts the digital number to BBE is 0.0001. QC is the quality control flag which gives a pixel-wise description of the data processing parameters as well as the credibility of the result. The flag is an 8-bit unsigned integer provided for each pixel. The bitwise interpretation of the QC flags for the GLASS BBE product is given in Table 4.12.

4.3.3 Validation

Field measurements are of vital importance for validating or testing the algorithms developed to retrieve specific biogeophysical parameters from satellite or aircraft data (Sobrino et al. 2006). During the past 6 years, a number of field experiments have been conducted in China in which the emissivities of large homogeneous surfaces were obtained (Buermann et al. 2001). These emissivity data were used to validate this new algorithm.

The first field experiment was carried out on August 18, 2006, at the Dunhuang Calibration Site in China for radiometric calibration of domestic satellites. Field measurement radiometers (CIMEL CE312-1) and an Infragold board were used to determine the emissivity of the Gobi Desert surface. The CIMEL CE312-1 was initially calibrated using a thermal infrared blackbody at five known temperatures, while the radiation values of the target and the environment were measured alternately. The measurements were repeated five times. The narrowband emissivity was determined using the ASTER TES algorithm (Gillespie et al. 1998). Following the same method documented in Sect. 4.2, the equation for converting the CE312 narrowband emissivities to the BBE values at 8–13.5 μm was derived. The average BBE value was 0.945, while the BBE value calculated using the new algorithm was 0.946, a difference of 0.001.

Two field experiments were then carried out on July 31, 2007 and October 12, 2007 at the Dunhuang Calibration Site in China. A Bomem MR 154 Fourier transform infrared (FTIR) spectrometer and a Labsphere gold plate were used to measure the spectral radiance emitted by the Gobi Desert surface and the environment. The measurements were repeated ten times for each target. The emissivity spectrum was derived from the radiometric measurements using the iterative spectrally smooth temperature and emissivity separation (ISSTES) algorithm (Borel 1998). The measured emissivity spectra were converted to the BBE at 8–13.5 μm. The average BBE measurements were 0.927 and 0.923, respectively, for the target and the environment, and the estimated BBEs were 0.947 and 0.945, giving an average difference of 0.021.

The final field experiment was conducted on June 6, 2011, in the Taklimakan Desert, Xinjiang Province, China, the largest desert in China, and the second largest in the world. A Model 102 Portable Field Spectrometer and a Labsphere gold plate were used to measure the spectral radiance emitted by the target and the environment. The emissivity spectrum was derived from the radiometric measurements using the ISSTES algorithm (Borel 1998). Two relatively homogeneous sites were selected in the central part of the Taklimakan Desert, and the three measurements at each site were made, and three points were randomly selected within a distance of approximately 500 m from the site where the three measurements were made at each point. The 12 emissivity spectra were averaged and regarded as the measured emissivity of the site. The average emissivity spectra were converted into the BBE values for the 8–13.5 μm spectral range. The measured BBE values were 0.915 and 0.913 for the target and the environment,

Table 4.13 Difference between the retrieved and the field-measured BBE values as well as the ASTER BBE values

Dune Sites	Field	ASTER	Retrieved	Retrieved (field)	Retrieved (ASTER)
Algodones	0.906	0.900	0.920	0.014	0.020
Great Sands	0.924	0.946	0.945	0.021	−0.001
Kelso	0.907	–	0.933	0.026	–
Little Sahara	0.914	0.945	0.942	0.028	−0.003
Stovepipe Wells	0.936	0.930	0.935	−0.001	0.005
White Sands	0.923	–	0.966	0.043	–

respectively, whereas the BBE values estimated using the new algorithm were 0.928 and 0.929, an average difference of 0.015. According to the results of the four field trials, the average difference between the estimated and the measured BBE values was 0.013.

Hulley and Hook (2009b) conducted five separate field trials to collect sand samples from nine sand dunes during the spring and early summer of 2008 to validate the North American ASTER Land Surface Emissivity Database (NA-ALSED). The emissivity spectra of the samples were measured in a laboratory

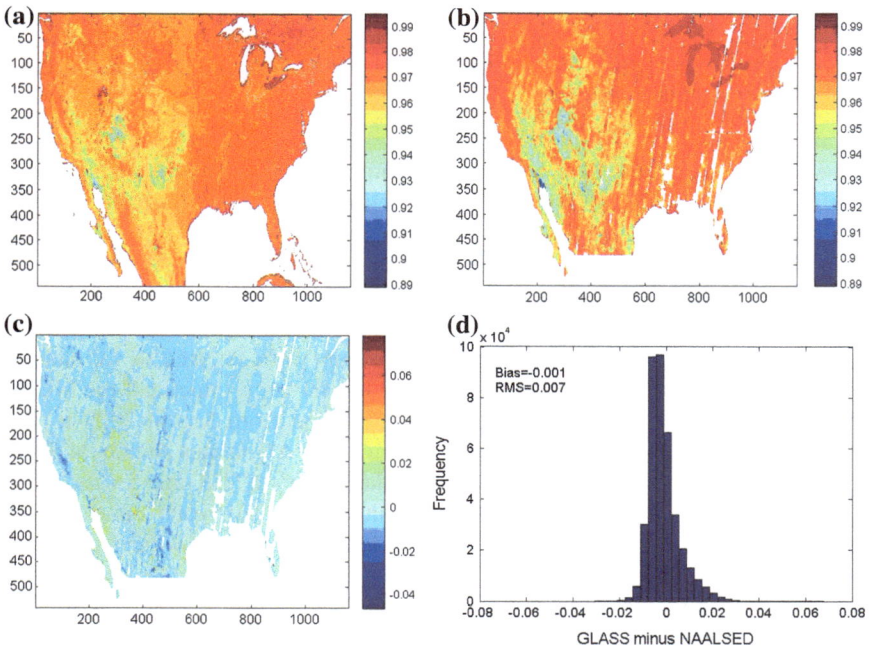

Fig. 4.37 Comparison between the NAALSED BBE and the GLASS BBE for the summer season. **a** GLASS BBE; **b** NAALSED BBE; **c** difference between the NAALSED BBE and the GLASS BBE; **d** histogram of the differences

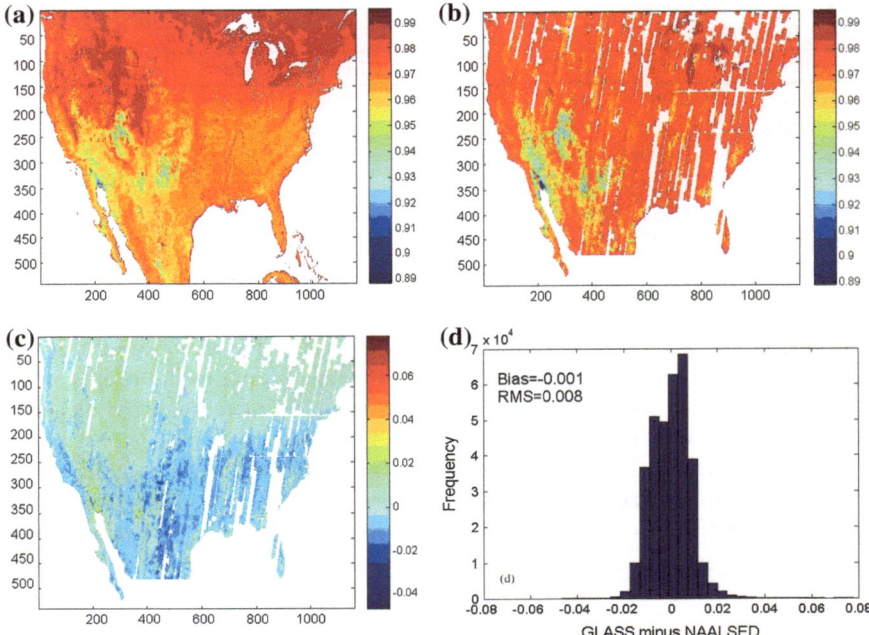

Fig. 4.38 Comparison between the NAALSED BBE and the GLASS BBE for the winter season. **a** GLASS BBE; **b** NAALSED BBE; **c** difference between the NAALSED BBE and the GLASS BBE; **d** histogram of the differences

using a Nicolet 520 FT-IR spectrometer equipped with a Labsphere integrating sphere (Baldridge et al. 2009). The corresponding ASTER narrowband emissivities were derived by convolving the laboratory-measured emissivity spectra with the ASTER TIR spectral response functions. The mineralogy of each dune site was also measured using X-ray diffraction (XRD). Further details on the dune sites can be found in the published paper.

Six relatively large and homogeneous dune sites were selected to validate this new algorithm. The spatially matched ASTER emissivity product and the MODIS albedo product were also downloaded for the period from March 2008 to June 2008. Overall, nine ASTER images were acquired for Algodones and Great Sands, eight ASTER images for Kelso and Little Sahara, ten ASTER images for Stovepipe Wells, and 34 ASTER images for White Sands. For each dune site, the derived narrowband emissivity and downloaded ASTER narrowband emissivity were converted to the BBE values in the 8–13.5 μm spectral range using Eq. (4.2) and these values were compared with those calculated using the new algorithm.

Table 4.13 shows the difference between the estimated and the measured BBE values, and that between the retrieved and the ASTER BBE values. The estimated BBE agreed well with the ASTER BBE with a difference of 0.005. The difference between the estimated and the measured BBE values was 0.022. For White Sands,

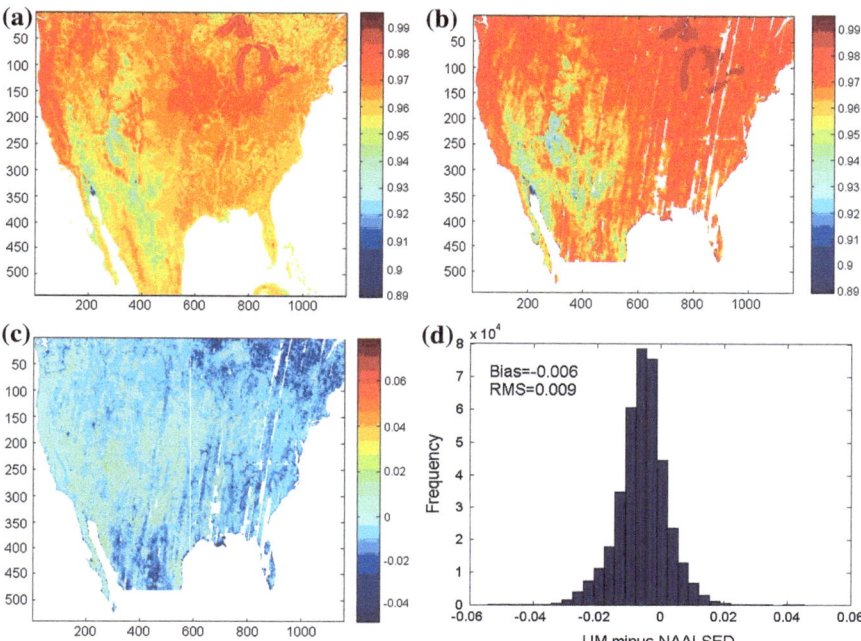

Fig. 4.39 Comparison between the NAALSED BBE and the UWIREMIS BBE for the summer season. **a** UWIREMIS BBE; **b** NAALSED BBE; **c** difference between the NAALSED BBE and the UWIREMIS BBE; **d** histogram of the differences

the BBE estimated using the new algorithm was 0.966, which was 0.043 higher than the measured BBE. The reason for this large difference was investigated. The corresponding soil order for White Sands was aridisol. The estimated narrowband albedos used to calculate the BBE were fairly large and clearly different from the average of the estimated albedos that was used to develop the algorithm.

A few extreme points also deviated from the 1:1 line in the scatter plots shown in Fig. 4.7. These extreme points were aridisols with narrowband albedos similar to those of White Sands. The established relationship did not represent these points well. The difference was 0.018 when this site was excluded. When combined with the validation results from our field measurements, the average difference between the estimated and the measured BBE values was 0.016.

4.4 Preliminary Analysis

The GLASS BBE was compared to the North American ASTER Land Surface Emissivity Database (NAALSED) and to the University of Wisconsin Global Infrared Land Surface Emissivity Database (UWIREMIS). The NAALSED is a

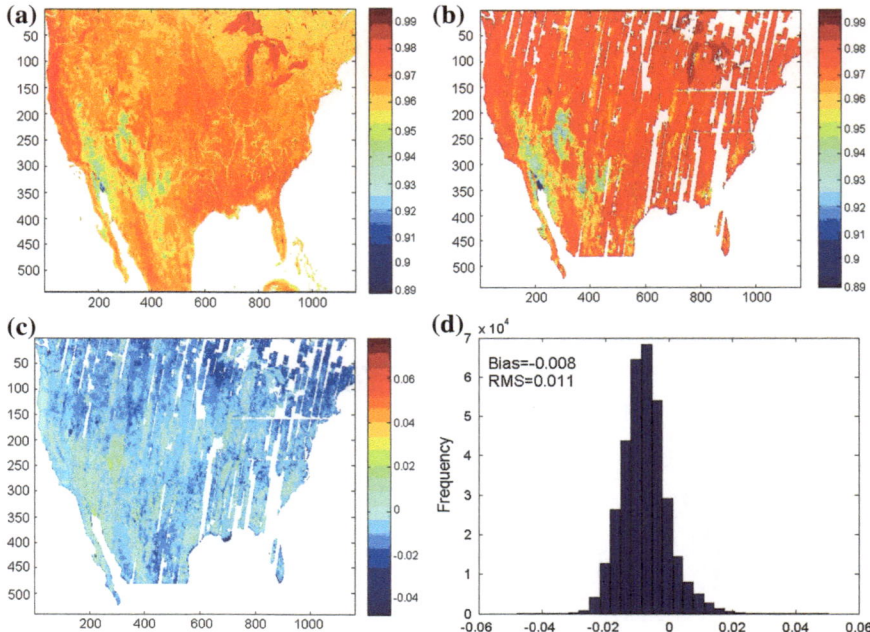

Fig. 4.40 Comparison between the NAALSED BBE and the UWIREMIS BBE for the winter season. **a** UWIREMIS BBE; **b** NAALSED BBE; **c** difference between the NAALSED BBE and the UWIREMIS BBE; **d** histogram of the differences

mean seasonal gridded 100 m emissivity database composed from ASTER 90 m standard land surface temperature and emissivity (LST&E) products over North America (Hulley and Hook 2009b). NAALSED includes two seasons, the summer season (July–September) and the winter season (Janunary–March). In the generation of NAALSED, cloud-contaminated ASTER pixels were screened out (Hulley and Hook 2008). For each location, the NAALSED emissivity is the average emissivity of all-clear-sky pixels from all ASTER scenes acquired in the summer and winter seasons from 2000 to 2008. NAALSED also produced the gridded emissivity products at spatial resolutions of 1, 5, and 50 km by aggregating the 100 m emissivity product. The NAALSED V2.0 product consists of eighteen bands: mean and standard deviation for the five bands of surface narrowband emissivity, surface temperature, NDVI, a land–water map, the total yield (number of ASTER observations collected at each pixel), and geodetic latitude and longitude. NAALSED was validated by the laboratory-measured sand emissivity collected at nine pseudo-invariant sand sites in the western United States (Hulley et al. 2009). The mean difference for all nine sites and all five ASTER channels were found to be 0.016, which represents approximately a 1 K error in LST retrieval.

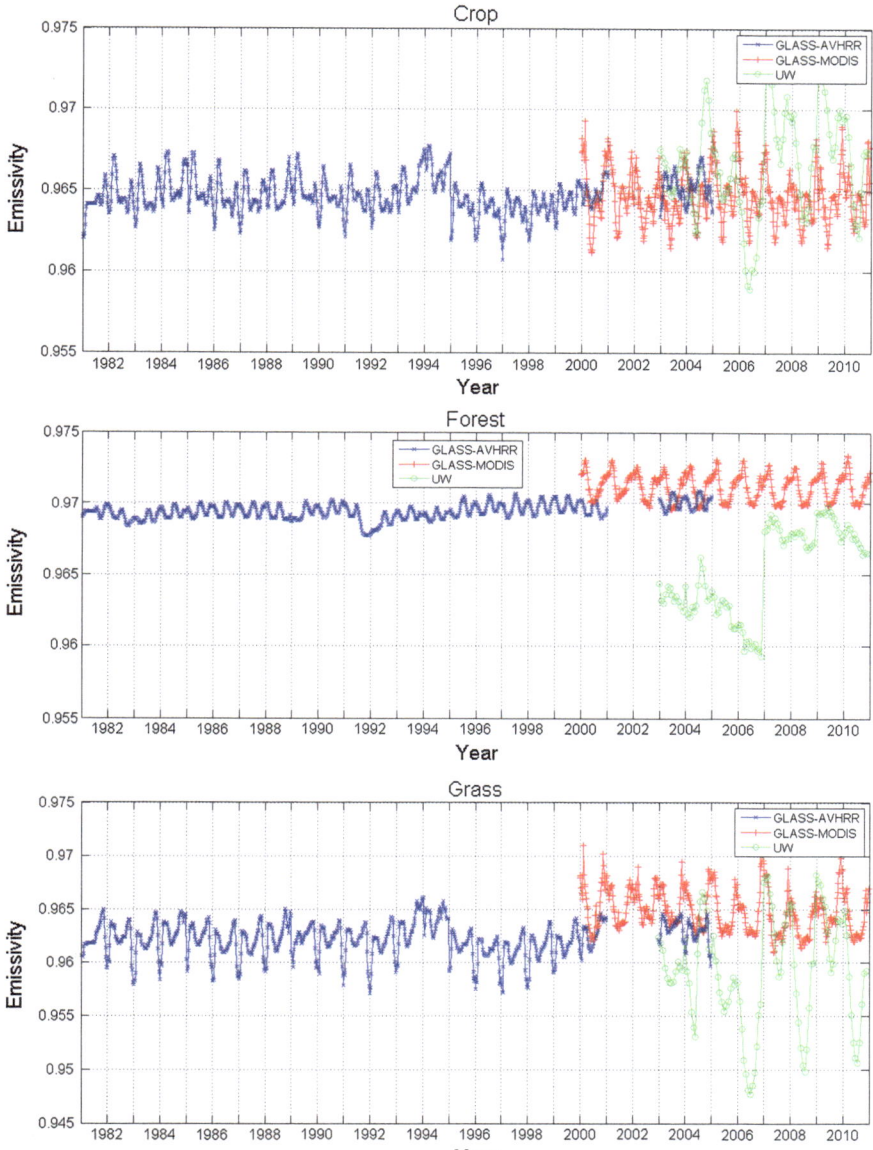

Fig. 4.41 Long-term global surface thermal broadband emissivity mean values of five land cover types from the GLASS emissivity product. The UW-Madison CIMSS emissivity product (converted from the narrowband emissivity values) is also used for comparison. The land cover maps are from the MODIS product: Majority_Land_Cover_Type_1 (MCD12C1 V5) and the land cover map for 2000 is used for the mean value calculation before 2000 (Liang 2013)

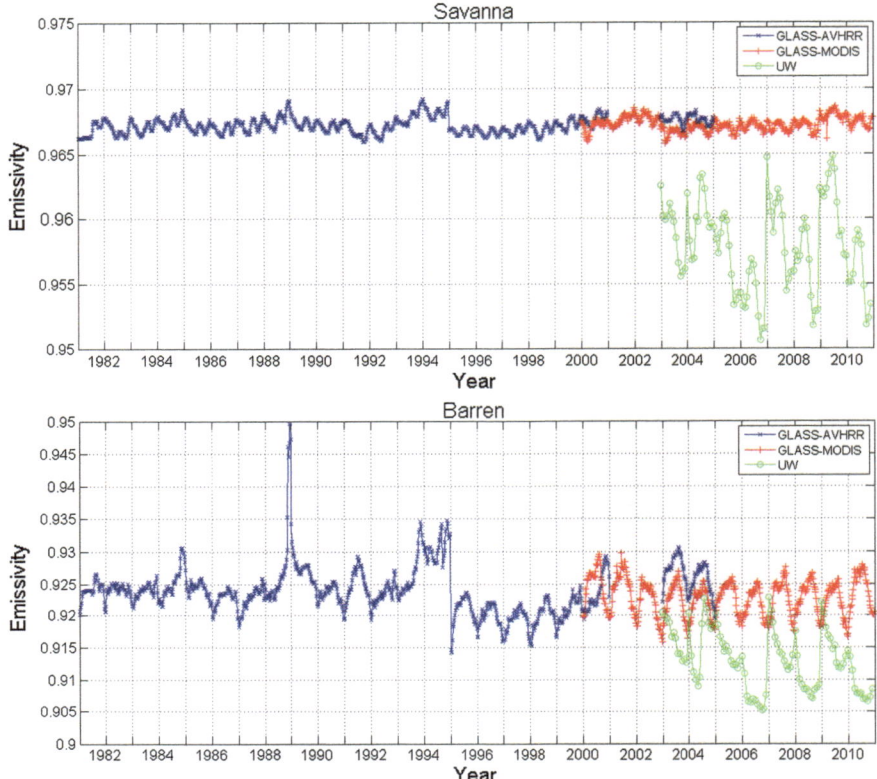

Fig. 4.41 continued

UWIREMIS is a monthly data set derived from the MODIS composited monthly 0.05° narrowband emissivity product (MOD11) by the baseline fit (BF) method (Seemann et al. 2008). The BF method derived global land surface emissivity at ten hinge points (3.6, 4.3, 5.0, 5.8, 7.6, 8.3, 9.3 10.8, 12.1, and 14.3 μm) by adjusting a baseline emissivity spectrum based on the MOD11 land surface narrowband emissivity product according to a conceptual model of land surface emissivity. Testing by the 123 emissivity spectra in the MODIS UCSB emissivity library indicated that the BF-derived emissivity generally agrees well with the laboratory-measured emissivity in shape and magnitude. The UWIREMIS emissivity could be interpolated between the hinge points and also could be used to derive a high spectral resolution emissivity spectrum by principal component regression and eigenvector from laboratory-measured emissivity spectra (Borbas et al. 2007).

NAALSED emissivity and UWIREMIS emissivity data were converted to the BBE at 8–13.5 μm, and the results were called NAALSED BBE and UWIREMIS BBE accordingly. Then, the NAALSED BBE was taken as the reference to evaluate the GLASS BBE and the UWIREMIS BBE in North America. The comparison results are shown in Figs. 4.37, 4.38, 4.39, and 4.40. The GLASS BBE is in good agreement with the NAALSED BBE for both summer and winter seasons. The bias and root mean square error (RMSE) are -0.001 and 0.007 for the summer season, -0.001 and 0.008 for the winter season, respectively. However, the UWIREMIS BBE and the NAALSED BBE differ substantially. The bias and the RMSE are -0.006 and 0.009 for the summer season, -0.008 and 0.011 for the winter season, respectively.

Figure 4.41 displays the temporal variations of thermal broadband emissivity for a few major land cover types from the homogeneous areas of global products. There are some minor disagreements of emissivity values from AVHRR and MODIS data, but overall the long-term values are stable and consistent. In comparison, the UW-Madison CIMSS emissivity values have much larger variations for most land cover types.

4.5 Summary

Satellite-derived land surface BBE values with high spatial and temporal resolutions are important for improving studies of land surface energy balance. In this study, the linear and nonlinear relationships between land surface broadband emissivity (BBE) and seven MODIS narrowband albedos and AVHRR channels 1 and 2 reflectances , which makes possible the generation of the long-time series of high spatial–temporal global land surface BBE product. With the developed algorithms, the global 8-day 1 and 5 km GLASS BBE product from 1981 to 2010 was developed.

Similarly to the NDVI threshold method (Sobrino et al. 2008), this new algorithm is a statistics-based algorithm. Although the remote sensing community prefers to develop and use physical-based algorithms, a statistics-based algorithm is essential for some specific applications. For example, it would be possible to derive land surface temperature from single band thermal infrared sensor like TM without the help of the NDVI threshold method for determining land surface emissivity in advance. This is also the case for estimate of global 8-day 1- and 5-km land surface BBE.

The developed algorithms have the usual merits of this type of algorithm such as simplicity, wide applicability, and acceptable accuracy. However, they also have one unavoidable drawback. The algorithm will lose efficacy if the inputs deviate greatly from the general behavior of samples used to establish the algorithm. The difficulties of finding bare soil pixels for some minor soil orders mean that the set of the extracted BBE–albedo pairs was limited. Thus, the resulting equations might not be representative even if the established equation could pass the confidence test. The results were greatly improved for soil orders with a high number of bare soil pixels. The spatial resolution of the soil taxonomy was approximately 0.033°, which was inconsistent with the 1 km spatial resolution of the MODIS products used and the resized ASTER emissivity product. In addition, errors in soil order classification may affect the established relationship for each soil order. To avoid classification errors, an effort was made to derive one equation to represent the relationship between the longwave BBE and shortwave albedo for the bare soil and transition zone. Unfortunately, an acceptable result could not be obtained for some minor soil orders. Finer soil taxonomy data will be incorporated when it is available.

In the absence of effective ground measurements at coarse spatial resolutions, only the algorithm for bare soils at 1 km spatial resolution could be validated by limited data. The algorithm for vegetated area was not validated by ground measurements. More extensive field research campaigns or more ground measurements are needed to obtain a wide range of surface emissivities, especially for vegetated canopies.

References

Baldridge AM, Hook SJ, Grove CI, Rivera G (2009) The ASTER spectral library version 2.0. Remote Sens Environ 113:711–715

Bonan GB, Oleson KW, Vertenstein M, Levis S, Zeng X, Dai Y, Dickinson RE, Yang Z (2002) The land surface climatology of the community land model coupled to the NCAR community climate model. J Clim 15:3123–3149

Borbas EE, Knuteson RO, Seemann SW, Weisz E, Moy L, Huang H-L (2007) A high spectral resolution global land surface infrared emissivity database. The Joint 2007 EUMETSAT meteorological satellite conference and the 15th satellite meteorology and oceanography conference, European Organisation for the Exploitation of Meteorological Satellites, Amsterdam

Borel CC (1998) Surface emissivity and temperature retrieval for a hyperspectral sensonr. In: Proceedings of IEEE conference on geoscience and remote sensing, pp 504–509

Bsaibes A, Courault D, Baret F, Weiss M, Olioso A, Jacob F, Hagolle O, Marloie O, Bertrand N, Desfond V, Kzemipour F (2009) Albedo and LAI estimates from FORMOSAT-2 data for crop monitoring. Remote Sens Environ 113:716–729

Buermann W, Dong J, Zeng X, Myneni RB, Dickinson RE (2001) Evaluation of the utility of satellite-based vegetation leaf area index data for climate simulations. J Clim 14:3536–3550

CEOS & WMO (2000) CEOS/WMO online database: Satellite system and requirements. In: The Committee on Earth Observation Satellites, The World Meteorological Organization

Chedin A, Scott NA, Wahiche C, Moulinier P (1985) The improved initialization inversion method: a high resolution physical mehtod for temperature retrievals from satellites of the TIROS-N series. J Climate Appl Meteorol 24:128–143

Cheng J, Liang S (2013a) Estimating the broadband longwave emissivity of global bare soil from the MODIS shortwave Albedo Product. J Geo Res Atmos (in press)

Cheng J, Liang S (2013b) Estimating global land surface broadband thermal-infrared emissivity from the Advanced Very High Resolution Radiometer optical data. Int J Digital Earth. doi:10.1080/17538947.2013.783129

Cheng J, Liu Q, Li X, Qing X, Liu Q, Du Y (2008) Correlation-based temperature and emissivity separation algorithm. Sci China, Ser D Earth Sci 51:363–372

Cheng J, Liang S, Wang J, Li X (2010a) A stepwise refining algorithm of temperature and emissivity separation for hyperspectral thermal infrared data. IEEE Trans Geosci Remote Sens 48:1588–1597

Cheng J, Liang S, Weng F, Wang J, Li X (2010b) Comparison of radiative transfer models for simulating snow surface thermal infrared emissivity. IEEE J Sel Top Earth Obs Remote Sens 3:323–336

Cheng J, Liang S, Liu Q, Li X (2011) Temperature and emissivity separation from ground-based MIR hyperspectral data. IEEE Trans Geosci Remote Sens 49:1473–1484

Dash P, Gottsche FM, Olesen FS, Fischer H (2002) Land surface temperature and emissivity estimation from passive sensor data: theory and practice—current trends. Int J Remote Sens 23:2563–2594

Gillespie AR, Rokugawa S, Matsunaga T, Cothern JS, Hook SJ, Kahle AB (1998) A temperature and emissivity separation algorithm for Advanced Spaceborne Thermal Emission and Reflection Radiometer (ASTER) images. IEEE Trans Geosci Remote Sens 36:1113–1126

Gillespie AR, Abbott EA, Gilson L, Hulley G, Jimenez-Munoz J-C, Sobrino JA (2011) Residual errors in ASTER temperature and emissivity products AST08 and AST05. Remote Sens Environ 115:3681–3694

Glotch TD, Rossman GR, Aharonson O (2007) Mid-infrared (5–100 um) reflectance spectra and optical constants of ten phyllosilicate minerals. Icarus 192:605–622

Gupta SK, Kratz DP, Wilber AC (2004) Validation of parameterized algorithms used to derive TRMM-CERES surface radiative fluxes. J Atmospheric Ocean Technol 21:742–752

Gustafson WT, Gillespie AR, Yamada G (2006) Revisions to the ASTER temperature/emissivity separation algorithm. In: Proceedings of the 2nd recent advance quantitative remote sensing, Valencia, Spain, pp 770–775

Hale GM, Querry MR (1973) Optical constants of water in the 200-nm to 200-um wavelength region. Appl Opt 12:555–563

Henning T, Il'In VB, Krivova NA, Michel B, Voshchinnikov NV (1999) WWW database of optical constants for astronomy. Astron Astrophys 136(Suppl):405–405

Hori M, Aoki T, Tanikawa T, Motoyoshi H, Hachikubo A, Sugiura K, Yasunari TJ, Eide H, Storvold R, Nakajima Y, Takahashi F (2006) In-situ measured spectral directional emissivity of snow and ice in the 8–14 um atmospheric window. Remote Sens Environ 100:486–502

Hulley GC, Hook SJ (2008) A new methodology for cloud detection and calssification with ASTER data. Geophys Res Lett 35:37

Hulley GC, Hook SJ (2009a) Intercomparison of versions 4, 4.1 and 5 of the MODIS land surface temperature and emissivity products and validation with laboratory measurements of sand samples from the Namib desert, Namibia. Remote Sens Environ 113:1313–1318

Hulley GC, Hook SJ (2009b) The North American ASTER Land Surface Emissivity Database (NAALSED) Version 2.0. Remote Sens Environ 113:1967–1975

Hulley GC, Hook SJ, Baldridge AM (2009) Validation of the North American ASTER Land Surface Emissivity Database (NAALSED) version 2.0 using pseudo-invariant sand dune sites. Remote Sens Environ 113(2224):2233

Jin M, Liang S (2006) An improved land surface emissivity parameter for land surface models using global remote sensing observations. J Clim 19:2867–2881

Li Z-L, Becker F (1993) Feasibility of land surface temperature and emissivity determination from AVHRR data. Remote Sens Environ 43:67–85

Li X, Strahler AH, Friedl MA (1999) A conceptual model for effective directional emissivity from nonisothermal surfaces. IEEE Trans Geosci Remote Sens 37:2508–2517

Li Z-L, Wu H, Wang N, Qiu S, SobrinoJA, Wan Z-M, Tang B-H, Yan G-J (2013) Land surface emissivity retrieval from satellite data. Int J Remote Sens 34:1–44

Liang S (2001) An optimization algorithm for separating land surface temperature and emissivity from multispectral thermal infrared imagery. IEEE Trans Geosci Remote Sens 39(264):274

Liang S (2004) Quantitative remote sensing of land surface. Wiley, Jew Jersey

Liang S, Wang K, Zhang X, Wild M (2010) Review on estimation of land surface radiation and energy budgets from ground measurement, remote sensing and model simulations. IEEE J Special Top Appl Earth Obs Remote Sens 3:225–240

Liang S, Zhao X, Liu S, Yuan W, Cheng X, Xiao Z, Zhang X, Liu Q, Cheng J, Tang H, Qu Y, Bo Y, Qu Y, Ren H, Yu K, Townshend J (2013) A long-term Global LAnd Surface Satellite (GLASS) data-set for environmental studies. Int J Digital Earth, doi:10.1080/17538947.17532013. 17805262

Liu Q, Xu X (1998) The retrieval of land surface temperature and emissivity by remote sensing data: theory and digital simulation. J Remote Sens 2:1–9

Masuda K, Takashima T, Takayama Y (1988) Emissivity of pure and sea waters for the model sea surface in the infrared window regions. Remote Sens Environ 24:313–329

Momeni M, Saradjian MR (2007) Evaluating NDVI-based emissivities of MODIS bands 31 and 32 using emissivities derived by Day/Night LST algorithm. Remote Sens Environ 106:190–198

Ogawa K, Schmugge T (2004) Mapping surface broadband emissivity of the sahara desert using ASTER and MODIS data. Earth Interact 8:1–14

Ogawa K, Schmugge T, Rokugawa S (2008) Estimating broadband emissivity of arid regions and its seasonal variations using thermal infrared remote sensing. IEEE Trans Geosci Remote Sens 46(334):343

Olioso A (1995) Simulating the relationship between thermal emissivity and the normalized difference vegetation index. Int J Remote Sens 16:3211–3216

Pedelty J, Devadiga S, MasuokaE, Brown M, Pinzon J, Tucker C, Roy D, Ju JC, Vermote E, Prince S, Nagol J, Justice C, Schaaf C, Liu JC, Privette J, Pinheiro A (2007) Generating a long-term land data record from the AVHRR and MODIS instruments. In: IEEE International geoscience and remote sensing symposium, pp 1021–1024. IEEE, New York

Pequignot E, Chedin A, Scott NA (2008). Infrared continental surface emissivity spectra retrieved from AIRS Hyperspectral sensor. J Appl Meteor Clim 47. doi:10.1175/2007JAMC1773.1

Pitman KM, wolff MJ, Clayton GC (2005) Application of modern radiative transfer tools to model laboratory quartz emissivity. J Geophys Res 110. doi:10.1029/2005JE002428

Ren H, Liang S, Yan G, Cheng J (2013) Empirical algorithms to map global broadband emissivities over vegetated surfaces. IEEE Trans Geosci Remote Sens 51:2619–2631

Seemann SW, Borbas EE, knuteson RO, Stephenson GR, Huang H-L (2008) Development of a global infrared land surface emissivity database for application to clear sky sounding retrieval from multispectral satellite radiance measurements. J Appl Meteor Clim 47:108–123

Sellers PJ, Dickinson RE, Randall DA, Betts AK, Hall FG, Berry JA, Collatz GJ, Denning AS, Mooney HA, Nobre CA, Sato N, Field CB, Henderson-Sellers A (1997) Modeling the exchange of energy, water and carbon between the continents and the atmosphere. Science 275:502–509

Sobrino JA, Raissouni N, Li Z-L (2001) A comparative study of land surface emissivity retrieval using NOAA data. Remote Sens Environ 75:256–266

Sobrino JA, Jimenez-Munoz JC, Zarco-Tejada PJ, Sepulcre-Canto G, Miguel ED (2006) Land surface temperature derived from airborne hyperspectral scanner thermal infrared data. Remote Sens Environ 102:99–115

Sobrino JA, Jiménez-Muñoz JC, Sòria G, Romaguera M, Guanter L, Moreno J, Plaza A, Martinez P (2008) Land surface emissivity retrieval from different VNIR and TIR sensors. IEEE Trans Geosci Remote Sens 46:316–327

Tang B-H, Wu H, Li C, Li Z-H (2011) Estimation of broadband surface emissivity from narrowband emissivities. Optical Express 19:185–192

Valor E, Caselles V (1996) Mapping land surface emissivity from NDVI: application to European, African, and South American areas. Remote Sens Environ 57:167–184

Van de Griend AA, Owe M, Groen M, Stoll MP (1991) Measurement and spatial variation of thermal infrared surface emissivity in a savanna environment. Water Resour Res 27:371–379

Wan Z, Li Z-L (1997) A Physics-based algorithm for retrieving land-surface emissivity and temperature from EOS/MODIS data. IEEE Trans Geosci Remote Sens 35:980–996

Wang K, Liang S (2009a) Evaluation of ASTER and MODIS land surface temperature and emissivity products usning long-term surface longwave radiation observations at SURFRAD sites. Remote Sens Environ 113:1556–1565

Wang W, Liang S (2009b) Estimation of high-spatial resolution clear-sky longwave downward and net radiation over land surfaces from MODIS data. Remote Sens Environ 113:745–754

Wang K, Wan Z, Wang P, Sparrow M, Liu J, Zhou X, Haginoya S (2005) Estimation of surface long wave radiation and broadband emissivity using Moderate Resolution Imaging Spectroradiometer (MODIS) land surface temperature/emissivity products. J Geophys Res 110:D11109. doi:10.1029/2004JD005566

Wang X, Ouyang XY, Tang B-H, Li Z-L, Zhang R (2008) A new method for temperature/emissivity separation from hyperspectral thermal infrared data. In: IGARSS'08, pp 286–289

Wang W, Liang S, Augustine JA (2009) Estimating high spatial resolution clear-sky land surface upwelling longwave radiation from MODIS data. IEEE Trans Geosci Remote Sens 47:1559–1570

Warren SG, Brandt RE (2008) Optical constants of ice from the ultraviolet to the microwave: a revised compilation. J Geophys Res 113. doi:10.1029/2007JD009744

Wilber AC, Kratz DP, Gupta SK (1999) Surface emissivity maps for use in satellite retrievals of longwave radiation. NASA Technical Publications, NASA/TP-1999-209362. http://techreports.larc.nasa.gov/ltrs

Yu Y, Tarpley D, Privette JL, Goldberg MD, Raja MKRV, Vinnikov KY, Xu H (2009) Developing algorithm for operational GOES-R land surface temperature product. IEEE Trans Geosci Remote Sens 47:936–951

Zhou L, Dickinson RE, Tian Y, Jin M, Ogawa K, Yu H, Schmugge T (2003) A sensitivity study of climate and energy blance simulations with use of satellite-based emissivity data over northern africa and the arabian peninsula. J Geophys Res 108:4795. doi:4710.1029/2003JD004083

Chapter 5
Incident Shortwave Radiation

Abstract Incident shortwave radiation (ISR), also known as insolation, is referred to as total solar irradiance incident at the Earth's surface and is an essential parameter in land surface radiation budget and many land surface process models. This chapter provides a primary introduction to the GLASS ISR product by discussing the algorithm, validation, and analysis. In the first section, a brief introduction will be given. The satellite data used and the implementation of the algorithms are discussed in Sect. 5.2. GLASS ISR product quality control and evaluation using ground measurements are presented in Sect. 5.3. The preliminary analysis and applications are described in Sect. 5.4, followed by a short summary.

Keywords Insolation · Shortwave radiation · Global irradiance · Remote sensing · MODIS · MSG · MTSAT · GOES

5.1 Background

The Sun is the only energy source for the climate system. Sunlight, which penetrates the atmosphere and reaches the Earth surfaces, is crucial for life on our planet. It not only causes the formation of cloud, fog, snow, and rain, but also heats the environment, induces pressure gradients, generates evaporation, and enables photosynthesis (Liang et al. 2012; Wild 2011). The incident shortwave solar radiation, also known as insolation, is scattered mainly by clouds, aerosols, or air molecules and absorbed by other gases. Thus, only part of it can penetrate the atmosphere and reach the Earth's surface in the form of direct and diffuse radiation.

Most solar radiation (99 % of its total) is concentrated within the spectral range of 250–2,500 nm. The visible, infrared, and ultraviolet bands contain about 50, 44, and 6 % of the total solar radiation, respectively. The incident shortwave radiation is mainly influenced by atmospheric properties and to a lesser extent by surface conditions (Chen et al. 2012). Clouds are considered to be the most important modulator of the solar radiant energy absorbed by the Earth-atmosphere system due to their strong interaction with solar radiation (Arking 1991). Thus, it is of great importance to generate accurate incident shortwave radiation product, especially under cloudy conditions.

Radiative transfer models (RTMs) are used to estimate or simulate radiation field under given atmospheric and surface conditions. To date, various RTMs have been reported, which can be divided into two categories: one-dimensional (1D) RTMs (Fu and Liou 1993; Ricchiazzi et al. 1998; Stamnes et al. 1988) and three-dimensional (3D) RTMs (Mayer et al. 2010; Mayer and Kylling 2005). The traditional (1D) RTMs, assuming homogeneous flat surfaces, may be not capable of simulating the topographic variations of radiative fluxes in three-dimensional terrain (Hoch and Whiteman 2010; Mayer et al. 2010) and they neglect radiative transfer effects from different atmospheric columns because they treat each pixel in a satellite image independently (Chen et al. 2012). It is more accurate to estimate the surface radiation budget using 3D RTMs. However, low computational efficiency makes them difficult to operate (Marshak and Davis 2005).

Besides RTM models, there are two ways to estimate surface incident shortwave radiation: surface data-based radiation models and satellite remote sensing models. Various surface data-based radiation models, which derive the surface incident shortwave radiation using sunshine duration and other meteorological parameters, have been developed (Jin et al. 2005; Paulescu and Schlett 2003; Psiloglou and Kambezidis 2007; Yang et al. 2010). Surface data-based radiation models commonly estimate incident shortwave radiation by establishing a relationship between ISR and sunshine duration or meteorological parameters, but the coefficients of those empirical methods are always site-dependent.

Use of satellite remote sensing is the most practical way of mapping incident shortwave radiation globally due to the large spatial coverage of satellite data (Liang 2004). The parameterization methods are one of the most widely used satellite remote sensing models for ISR estimation (Gueymard 2003; Ryu et al. 2008; Van Laake and Sanchez-Azofeifa 2004; Yang et al. 2010). These parameterization methods estimate surface solar radiation by calculating the amount of radiation absorbed, reflected and scattered by water vapor, gas, ozone, and aerosols (Zhang and Liang 2012). However, Gui et al. (2010) showed that satellite-based downward surface shortwave irradiance datasets (the Water-Cycle Experiment (GEWEX) SRB, the International Satellite Cloud Climatology Project (ISCCP) FD, and the Clouds and Earth's Radiant Energy System (CERES) FSW) had large biases in Southeast Asia, the Tibetan Plateau, and Greenland. Discrepancies among satellite-derived shortwave radiation products are usually larger in highly variable terrain and smaller for non-variable terrain (Yang et al. 2010). Moreover, the existing satellite derived incident radiation products have low spatial resolution. For instance, the spatial resolution of the GEWEX SRB dataset is 1°, although the maximum temporal resolution is 3 h. Thus, additional radiation algorithms with greater accuracy should be developed.

This chapter will provide a detailed introduction to the algorithm, validation, and analysis of the GLASS ISR product, which has a spatial and temporal resolution of

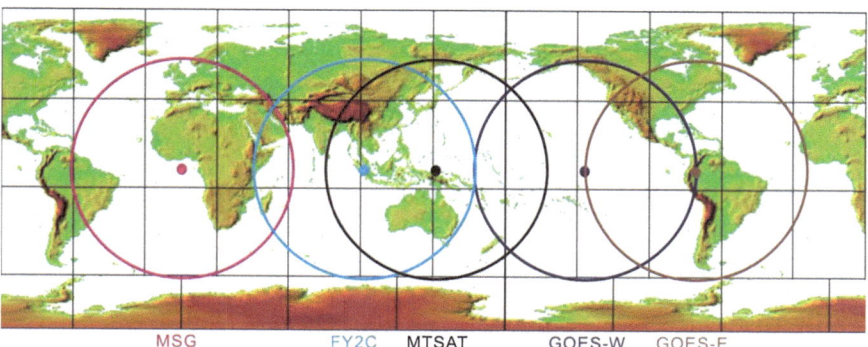

Fig. 5.1 Spatial coverage of the geostationary satellite used in the GLASS ISR product

5 km and 3 h, respectively. The GLASS ISR product was generated from both polar-orbiting and geostationary satellite data to achieve global coverage. Geostationary satellite data have a fine temporal resolution but small spatial coverage. On the other hand, polar-orbiting satellite data have fine spatial coverage but coarse temporal resolution. Thus, geostationary satellite data can be used to capture the diurnal variations of solar radiation, which is good for calculating the daily integrated incident shortwave radiation (Laszlo et al. 2008). The radiation estimation precision will be improved if both polar-orbiting and geostationary satellite data are used simultaneously. The satellite data used in the GLASS ISR product include the Moderate-Resolution Imaging Spectroradiometer (MODIS), Meteosat Second Generation (MSG) SEVIRI, Multifunctional Transport Satellite (MTSAT)-1R, and Geostationary Operational Environmental Satellite (GOES) Imager. The spatial coverage of the geostationary satellite used is shown in Fig. 5.1.

The GLASS ISR product is generated using an improved look-up table approach using the top-of-atmosphere (TOA) radiance of the visible band for each sensor. The ozone and water vapor absorption, surface elevation, and Bidirectional Reflectance Distribution Function (BRDF) effects are taken into account in the algorithm. Detailed information on the GLASS ISR product will be provided in the following section. The data sources used in the GLASS ISR product and the algorithm itself are given in Sect. 5.2. The product characteristics, quality control, and evaluation are presented in Sect. 5.3. Section 5.4 presents a preliminary analysis and applications of the GLASS ISR product. A summary is provided at the end of this chapter.

5.2 Algorithms

The main idea of the GLASS ISR product is to try to establish the relationship between ISR and TOA radiance through radiative transfer simulation using the Moderate Resolution Transmission (MODTRAN) radiative transfer software package (Anderson et al. 1999). In this section, the data sources used in the GLASS ISR product and detailed information on GLASS ISR algorithm will be presented.

5.2.1 Data Sources

5.2.1.1 Polar-Orbiting Satellite Data

The MODIS sensors onboard the Terra and Aqua satellites have 36 spectral bands ranging from the visible to the thermal-infrared spectrum. The spatial resolution varies from 250 (bands 1 and 2) to 500 m (bands 3–7) and 1,000 m (band 8–36). The MODIS sensors view the entire Earth every 1 or 2 days with one or two passes or more at high latitudes per day. The blue band (band 3) is used for ISR estimation.

The MODIS geolocation dataset (MOD03 and MYD03), land surface reflectance product (MOD09A1) (Vermote et al. 2002), BRDF/albedo parameter product (MCD43B1) (Schaaf et al. 2002), and the precipitable water product (MOD08_D3) (Hubanks et al. 2008) were also used in the proposed algorithm. The geolocation dataset provides geodetic coordinates, ground elevation, solar and satellite zenith, and azimuth angle for each MODIS 1 km sample. The land surface reflectance products (MOD09A1) provide an estimate of the surface spectral reflectance as it would be measured at the ground level in the absence of atmospheric scattering or absorption of MODIS bands 1–7 at 500 m resolution and an day 8 gridded level-3 product. Each MOD09A1 pixel contains the best-possible L2G observation during an 8-day period as selected based on high observation, low view angle, the absence of clouds or cloud shadow, and aerosol loadings.

The MODIS BRDF/albedo parameter product (MCD43B1) provides the BRDF model parameters with which the user can reconstruct the surface and the BRDF as well as estimate the directional reflectance at any view or solar zenith angle. The directional hemisphere reflectance and the bi-hemisphere reflectance can be estimated using these parameters at any desired viewing angle or solar zenith angle (Schaaf et al. 2002). The MODIS atmospheric precipitable water product (MOD08_D3) estimates the total water vapor column over clear land surfaces globally and above clouds over both land and sea. The near-infrared total precipitable water column product, which is essential for understanding the hydrological cycle, aerosol properties, aerosol-cloud interactions, energy budget, and climate, was used in this study. Under cloudy conditions, the water vapor column above the clouds was used instead of the total water vapor amounts because the influence of the amount of water vapor is not dominant under cloudy conditions. However, the influence of water vapor is significant under clear-sky conditions. To combine the MODIS satellite data with other geostationary satellite data, the original MODIS L1B swath data were converted into a sinusoidal projection with 5 km spatial resolution.

5.2.1.2 Geostationary Satellite Data

GOES11 is designated as GOES-West and is located at 135°W over the Pacific Ocean, whereas GOES12 is designated as GOES-East and is located at 75°W over the Amazon River. The imaging instrument on the current GOES has one visible band and four bands in the infrared spectrum. The GOES satellite has a distinct advantage over a polar-orbiting satellite in its ability to provide observation data with high spatial (≥ 1 km) and temporal resolutions (≥ 15 min) (Otkin et al. 2005). The GOES visible band data were used for GLASS ISR estimation.

The main payload of the MSG satellite is the optical imaging radiometer, which is called the Spinning Enhanced Visible and Infrared Imager (SEVIRI). This device collects images of the Earth's surface and atmosphere at different wavelengths once every 15 min, compared to three wavelengths once every 30 min for the comparable instrument on Meteosat. The MSG SEVIRI provides images with a resolution as low as 1 km in the visible band, which is sharper than the 2.5 km resolution of Meteosat. In the infrared, the SEVIRI resolves features that are 3 km across compared to 5 km for Meteosat.

The Multifunctional Transport Satellite (MTSAT) series succeeds the Geostationary Meteorology Satellite (GMS) series and fulfills meteorological functions (such as weather forecasts, natural-disaster countermeasures, and securing safe transportation) for the Japan Meteorology Agency. The MTSAT series satellites are located in a geostationary orbit 35,800 km above the equator at 140°E and 145°E. The MTSAT series carries a new imager with a new infrared channel (IR4) in addition to the four channels (VIS, IR1, IR2, IR3) of the GMS-5. The spatial resolution is 1 km (VIS) and 4 km (IR) at the subsatellite point.

5.2.1.3 Ancillary Data

Ground-measured pyranometer data were used to evaluate the GLASS ISR product. The elevation data were incorporated into the GLASS ISR data to improve the retrieved ISR for highly variable terrain, for instance, the Tibetan Plateau. The elevation data were downloaded from GTOPO30 at a spatial resolution of 30 arc s (http://eros.usgs.gov/#/Find_Data/Products_and_Data_Available/gtopo30_info).

5.2.2 Algorithm Description

5.2.2.1 Creating Look-Up Tables

The GLASS ISR product was generated using an improved look-up table method proposed by Liang et al. (2006) for photosynthetically active radiation (PAR) estimation. The look-up table was generated by atmospheric radiative transfer using MODTRAN simulation, which was done with the selected aerosol model and cloud type at different view geometries for each sensor with its own channel response function. The ISR under different atmospheric and surface properties at different geometries can be obtained by integrating the downward spectral flux from 300 to 3,000 nm resulting from MODTRAN simulation. The main procedure of the proposed algorithm is illustrated in Fig. 5.2.

Assuming that the surface exhibits Lambertian reflectance and that the atmosphere is horizontally uniform, the TOA apparent reflectance ρ_{TOA} and the downward spectral flux $F(\theta_s)$ at the surface for a specified solar zenith angle θ_s and a view zenith angle θ_v can be expressed by the following equations (Liang et al. 2006; Vermote et al. 1997, 2002):

$$\rho_{TOA} = \rho_a(\theta_s, \theta_v, \phi_s - \phi_v) + \frac{\rho_t(\theta_s, \theta_v, \phi_s - \phi_v)}{1 - \rho_t(\theta_s, \theta_v, \phi_s - \phi_v)S} T(\theta_s)T(\theta_v) \qquad (5.1)$$

$$F(\theta_s) = F_0(\theta_s) + \frac{\rho_t(\theta_s, \theta_v, \phi_s - \phi_v)S}{1 - \rho_t(\theta_s, \theta_v, \phi_s - \phi_v)S} \mu_s E_0 T(\theta_s) \qquad (5.2)$$

where $\rho_a(\theta_s, \theta_v, \phi_s - \phi_v)$ is the intrinsic reflectance of the atmosphere without any contribution from the surface, $\rho_t(\theta_s, \theta_v, \phi_s - \phi_v)$ is the surface reflectance, S is the spherical albedo of the atmosphere, $T(\theta_s)$ is the total transmittance from the TOA to the ground in the solar direction, $T(\theta_v)$ is the transmittance from the ground to the sensor in the viewing direction of the satellite, $F(\theta_s)$ is the surface downward flux at the specified solar zenith angle, $F_0(\theta_s)$ is the downward flux without any contribution from the surface, μ_s is the cosine of the solar zenith angle θ_s, and E_0 is the extraterrestrial solar irradiance.

At the satellite level, three different surface reflectances are specified to solve for $\rho_a(\theta_s, \theta_v, \phi_s - \phi_v)$, $T(\theta_s)T(\theta_v)$, and the spherical albedo S based on Eq. (5.1). The first look-up table is generated using these three parameters under specified atmospheric conditions and geometries. Thus, if these three variables are known, the relationship between the TOA apparent reflectance and the surface reflectance can be determined based on Eq. (5.1) under given atmospheric conditions and geometries.

Similarly, three different surface reflectances are used to solve for $F_0(\theta_s)$, the spherical albedo S, and $\mu_s E_0 T(\theta_s)$ to establish the second look-up table. It is also possible to connect the surface spectral flux using these three parameters and the surface reflectance using Eq. (5.2). Cloud-free and cloudy conditions are separated in the simulation but are listed together in the look-up table. Under the clear-sky conditions, the rural aerosol model is selected during the simulation by MODTRAN4. However, under cloudy conditions, altostratus cloud was used under the specified aerosol loadings (23 km) because the influence of aerosols can be neglected under cloudy conditions, but is significant under cloud-free conditions. The parameters set in MODTRAN4 used to create the two look-up tables described above is summarized in Table 5.1.

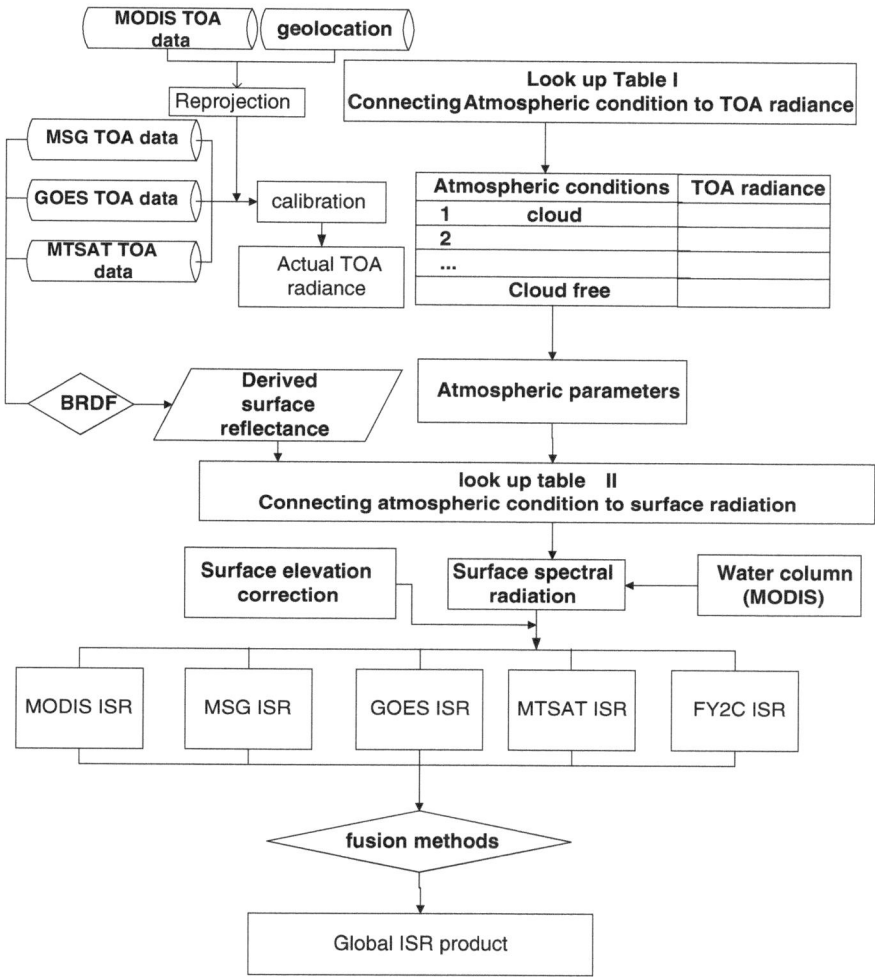

Fig. 5.2 Flow diagram of GLASS ISR estimation

5.2.2.2 Determination of Surface Reflectance

The determination of surface reflectance is one of the most important tasks in estimating ISR using the look-up table method. Because the characteristics are different for different sensors or satellites, the retrieval of surface reflectance was divided into two steps. The MODIS land surface reflectance product (MOD09A1) was used to provide input parameters for the MODIS sensor. However, for geostationary satellites, the minimum TOA blue band reflectance method was employed to derive the surface reflectance (Liang et al. 2006). Although the MODIS MOD09A1 product provides the surface reflectance in an 8-day period in

Table 5.1 Summary of essential parameters in the look-up table

Input parameters	Values
Solar zenith angle	0°, 15°, 30°, 45°, 55°, 65°, 75, 85°, 90°
View zenith angle	0°, 20°, 40°, 60°, 80°
Relative azimuth angle	0°, 30°, 60°, 90°, 120°, 150°, 180°
Surface altitude (km)	0.000, 1.500, 3.000, 4.500, 5.900
Aerosol type	Rural aerosol
Cloud type	Altostratus cloud
Aerosol loadings (visibility) (km)	5, 10, 20, 30, 100
Cloud optical thickness (km^{-1})	1, 2, 3, 5, 10, 25, 50, 70, 128
Water vapor amount	Default value
Ozone amount	Default value

the absence of clouds or cloud shadows and aerosol loadings, the surface reflectance for the MODIS sensor data remain contaminated by clouds. Thus, the MODIS 8-day composite land surface reflectance blue-band data were processed using a simple method to reduce the influences of clouds and snow, and the processed data were then utilized to retrieve ISR.

For the geostationary satellite, the BRDF effects were taken into account for the minimum TOA blue band using the MODIS BRDF/albedo parameter product (MCD43B1). The algorithm used here is the "clearest" observation method proposed by Liang et al. (2006). The operational MODIS BRDF/albedo product (MCD43B1) was then used to reduce the BRDF effect. The MODIS MCD43B1 BRDF product is a kernel-driven, linear BRDF model that is a weighted function of the isotropic parameter and two kernels for viewing and illumination geometry (Schaaf et al. 2002). The weighted function is presented in Eq. (3.2).

The $k_{vol}(\theta_s, \theta_v, \varphi)$ and $k_{geo}(\theta_s, \theta_v, \varphi)$ kernels can be derived using the following equations:

$$k_{vol}(\theta_s, \theta_v, \varphi) = \frac{(\pi/2 - \xi)\ \cos\ \xi + \ \sin\ \xi}{\cos\ \theta_s + \cos\ \theta_v} - \pi/4 \qquad (5.3)$$

$$k_{geo}(\theta_s, \theta_v, \varphi) = O(\theta_s, \theta_v, \phi_s - \phi_v) - \sec\theta'_s - \sec\theta'_v + \frac{1}{2}(1 + \cos\xi')\sec\ \theta'_s\sec\theta'_v \qquad (5.4)$$

$$\cos\xi = \cos\ \theta_s\cos\theta_v + \sin\theta_s\sin\theta_v\cos(\phi) \qquad (5.5)$$

$$O(\theta_s, \theta_v, \varphi) = \frac{1}{\pi}(t - \sin t\cos t)(\sec\theta'_s + \sec\theta'_v) \qquad (5.6)$$

$$\cos t = \frac{h}{b}\sqrt{\frac{D^2 + (\tan\theta'_s\tan\theta'_v\sin\phi)^2}{\sec\theta'_s + \sec\theta'_v}} \qquad (5.7)$$

$$D = \sqrt{\tan^2 \theta'_s + \tan^2 \theta'_v - 2\tan \theta'_s \tan \theta'_v \cos \phi} \tag{5.8}$$

$$\cos \xi' = \cos \theta'_s \cos \theta'_v + \sin \theta'_s \sin \theta'_v \cos(\phi)' \tag{5.9}$$

$$\theta'_s = \tan^{-1}\left(\frac{b}{r}\tan \theta_s\right) \text{ and } \theta'_v = \tan^{-1}\left(\frac{b}{r}\tan \theta_v\right) \tag{5.10}$$

where ξ is the phase angle, $O(\theta_s, \theta_v, \varphi)$ is the area of overlap between the view and solar shadows, and h/b and b/r are the crown shape parameters, which are equal to 2 and 1, respectively.

After the surface reflectance on the clearest days for different sensors has been obtained, these surface reflectance values can be employed to estimate the BRDF parameters. The surface reflectance can then be calculated using these BRDF parameters at the specified solar and view zenith angles.

5.2.2.3 Searching the LUT to Estimate ISR

The surface ISR values can be obtained using the created LUTs and the derived surface reflectance from the MODIS product. The main procedure is as follows:

First, the TOA reflectance and radiance for different sensors is calculated based on the DN numbers using the calibration coefficients. Second, the surface reflectance for different sensors is derived. Then the TOA radiance is estimated for each atmospheric condition, from the clearest to the cloudiest (high cloud extinction coefficient), based on the first LUT. Then, the actual TOA radiance calculated from the sensors is compared to the series of the simulated radiance values for different atmospheric conditions to retrieve the atmospheric index. Finally, incident shortwave radiation can be estimated by searching the second LUT using the estimated atmospheric condition index and surface reflectance.

5.2.2.4 Water Vapor Correction

The solar irradiances are reduced within the atmosphere. Water vapor absorbs infrared radiation, and the influence of water vapor on estimation of surface radiation is significant, especially in the infrared band under cloud-free conditions. Water vapor broadband and spectral transmission parameterization functions have been presented in the literature (Gueymard 2003; Ryu et al. 2008; Van Laake and Sanchez-Azofeifa 2004).

To improve the computational efficiency of atmospheric radiative transfer using MODTRAN, the water vapor amount was set to the default value. To quantify the influences of water vapor in the infrared band on ISR estimation, a simple normalized water vapor transmittance method was used for water vapor correction of the solar radiation at the surface. The water vapor transmission rates index can be estimated using the following equation (Psiloglou et al. 2000).

$$T_w = 1 - \frac{3.014Mu}{((1 + 119.3Mu)^{0.644} + 5.814Mu)} \qquad (5.11)$$

where M is the optical atmospheric mass, and u is the water vapor amount in cm. The optical air mass M can be obtained as:

$$M = (\cos(\theta_s) + 0.50572(96.07995 - \theta_s)^{-1.6364})^{-1} \qquad (5.12)$$

where θ_s is the solar zenith angle. After the water vapor transmittance has been calculated, the water vapor transmittance can be normalized by the following equation:

$$R_w = T_w(u)/T_w(u_d) \qquad (5.13)$$

where R_w is the normalized water vapor transmittance, $T_w(u)$ is the water vapor transmittance for the water vapor amount u in cm, and $T_w(u_d)$ is the water vapor transmittance for the default amount of water vapor.

5.2.2.5 Combination of ISR from Polar-Orbiting and Geostationary Satellites

To map ISR globally, both polar-orbiting and geostationary satellite data were used for GLASS ISR estimation. Geostationary satellites have a finer temporal resolution than MODIS, but are less reliable at high latitudes due to their large view angle. Thus, the MODIS data were used to derive the surface solar radiation at the north and south latitudes greater than 60°, and ISR at lower latitudes was calculated using a combination of polar-orbiting and geostationary satellite-derived radiation products.

The predicted surface solar radiation from different sensors is expressed by the variable r, and the corresponding in situ measurement at a given time point is r_t. The estimated ISR from multiple satellite data can be derived using the normalized posterior probability weighted value for each sensor. The weighted coefficient $p(f_k|r_t)$ is the posterior probability of the derived value for each sensor f_k, which can reflect how well the model f_k fits the observed data and is assumed to have a Gaussian distribution.

Because MODIS observes the surface one or more times per day at high latitudes, the look-up table method, proposed by Wang and Liang (2010) was used to interpolate for the time that is not observed during the day at high latitudes to fulfill the required 3 h temporal resolution of this dataset. This method assumes that the atmospheric condition varies little during a given period of time. The incident shortwave radiation was then estimated using the closest time derived from the satellite using the atmospheric index, estimated from the look-up table, under the same atmospheric conditions.

5.3 Product Characteristics, Quality Control, and Validation

5.3.1 Product Characteristics

The GLASS ISR product was generated using the algorithms and procedures described above using the GLASS production system, which primarily includes processing, management, and database servers. The processing servers generate the products using a High-Performance Computing (HPC) system. The management servers and database servers provide system operations management and data storage. The software of the production system generally includes production task management, HPC distribution, quality inspection, and system monitoring. The global GLASS ISR was generated from the estimated ISR of the selected polar-orbiting and geostationary satellite. This means that the GLASS ISR product includes not only the global fusioned product, but also that for each satellite. The spatial and temporal resolutions are summarized in Table 5.2.

5.3.2 Quality Control

The quality of the GLASS ISR product was primarily dependent on the sky conditions, snow cover, and surface reflectance. Thus, the quality of these two products is represented by a scientific dataset, which includes information on cloud state, cloud shadow, surface reflectance quality, and snow cover within a 16-bit unsigned integer. Detailed description of the GLASS ISR QC information is given in Table 5.3.

5.3.3 Validation

The validation experiments were first performed using publicly available data, which include one SURFRAD site, two AERONET sites, and one CarbonEuropeIP

Table 5.2 Spatial and temporal resolutions of GLASS ISR/PAR products

Product type (ISR/PAR)	Spatial resolution (km)	Temporal resolution (h)	Temporal coverage
Global	5	3	January 2008–December 2010
MSG	5	0.25	
MTSAT	5	1	
GOES-E	5	3	
GOES-W	5	3	

Table 5.3 GLASS ISR product QC description (16 bits)

Bit No.	Parameter name	Bit comb.	Quality control
1–2	Cloud state	00	Clear
		01	Cloudy
		10	Mixed
		11	Not set, assumed clear
3	Cloud shadow	1	Yes
		0	No
4–7	MODIS QC surface reflectance	0000	Highest quality
		1000	Dead detector; data interpolated in L1B
		1001	Solar zenith >= 86°
		1010	Solar zenith >= 85 and <86°
		1011	Missing input
		1100	Internal constant used in place of climatological data for at least one atmospheric constant
		1101	Correction out of bounds pixel constrained to extreme allowable value
		1110	L1B data faulty
		1111	Not processed due to deep ocean or clouds
8	Snow cover	1	Yes
		0	No
9	Surface reflectance process state	1	Original value
		0	Filter value
10–11	Surface reflectance QC	00	Very good
		01	Good
		10	Bad
		11	Unknown
12–16	Unused		

site. ISR ground measurements were collected every 3, 2, and 30 min at the SURFRAD site, the AERONET site, and the CarbonEuropeIP site, respectively. For the 2 and 30 min observation frequency sites, the respective 30 and 60 min averages that were closest to the time of derivation from satellite data were compared with the inverted values to reduce possible mismatches in space and time.

The first validation site was Goodwin Creek in Mississippi, which is located on rural pasture land with sparsely distributed deciduous trees. The tower is located at 34.25°N, 89.87°W. Comparison of the retrieved ISR values for 2008 shows a good fit (Fig. 5.3a). The R-squared value is approximately 0.9. The bias and RMSE for the retrieved ISR are 25.8 and 95.4 W m^{-2}, respectively.

The second site, Alta_Floresta (9.87°S, 56.1°W), is located in Alta Floresta, Brazil, at 277 m above sea level. The landscape is composed of grassland and sparse trees. The regional climate is warm, humid tropical with an annual temperature range from 22 to 37 °C. These conditions can cause the majority of days in this area to be hazy or cloudy, and therefore the retrieved solar radiation in this area may be affected. This site provided only measured ISR data in 2008. The

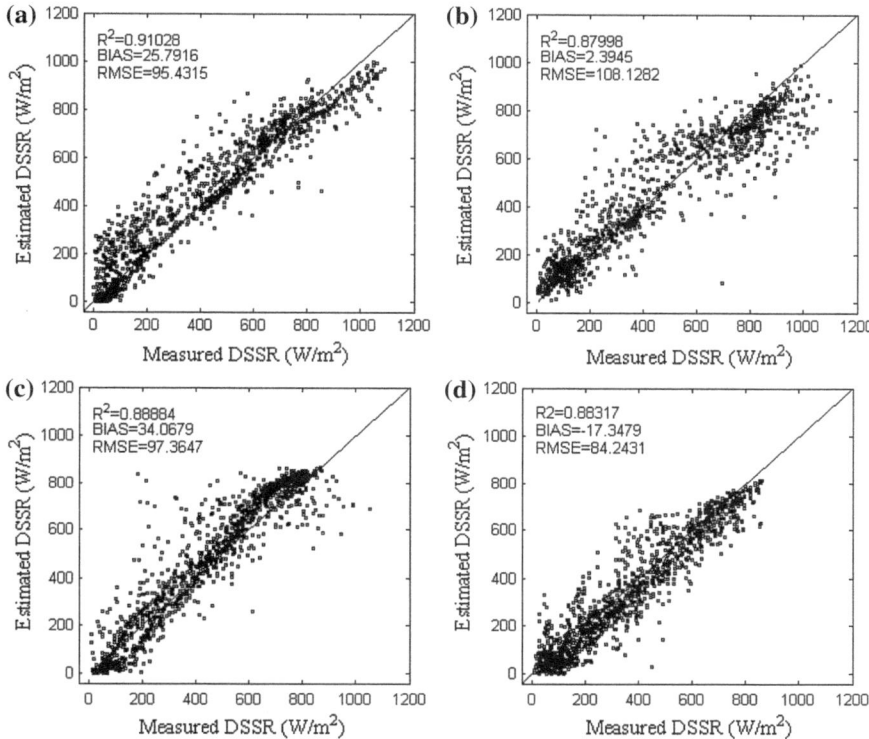

Fig. 5.3 GLASS ISR product validation results at the Goodwin Creek site (**a**) of SURFARAD, Alta_Floresta (**b**), and Moldova (**c**) from ARONET and the CZ-wet Site (**d**) from CarbonEuropeIP

overall fit to the ISR results at this site was good with occasional scattered points, as shown in Fig. 5.3b.

The third site was in Moldova (Kishinev, Moldova). The sun-photometer site at Kishinev is located on the roof of the Institute of Applied Physics building (20 m high). The geographic coordinates of this site are 47.0°N and 28.8°E. The climate in Moldova is temperate continental. The summers are warm and long, and the winters are relatively mild and dry. Measurements from 2008 were used, and the validation results are shown in Fig. 5.3c.

The last site was CZ-wet, which is located at 425 m above the sea level in the Czech Republic. The geographic coordinate of this site is 49.03°N, 14.77°E. The CZ-wet land surface is mostly covered by croplands. The ISR validation results at CZ-wet were good, as shown in Fig. 5.3d. The R-squared value is approximately 0.88.

Besides the validation experiments at publicly available sites, Huang et al. (2013) validated the instantaneous GLASS insolation product using ground measurements at 22 sites in arid and semi-arid regions of China and found that the R^2

at every site except one was greater than 0.8 and that the RMSE values ranged from 90 to 130 Wm^{-2}.

5.4 Preliminary Analysis

5.4.1 Global Mapping

In this study, both global and sensor-independent ISR products have been generated. The respective spatial and temporal resolutions are 5 km and 3 h for GOES11 and GOES12, 5 km and 1 h for MTSAT, and 5 km and 15 min for MSG2, as illustrated in Figs. 5.4, 5.5, 5.6, and 5.7. The globally fused products of the ISR images are mapped in a sinusoidal projection with a 5 km spatial resolution and a 3 h temporal resolution, as shown in Fig. 5.8.

5.4.2 Comparison with Other Products

The validation results after aggregation of the spatial resolutions of the GLASS insolation product to match the ISCCP and CERES products are shown in Table 5.4. The CERES surface shortwave Model B depends on simple relationships to estimate the attenuation of surface insolation by the Earth's atmosphere

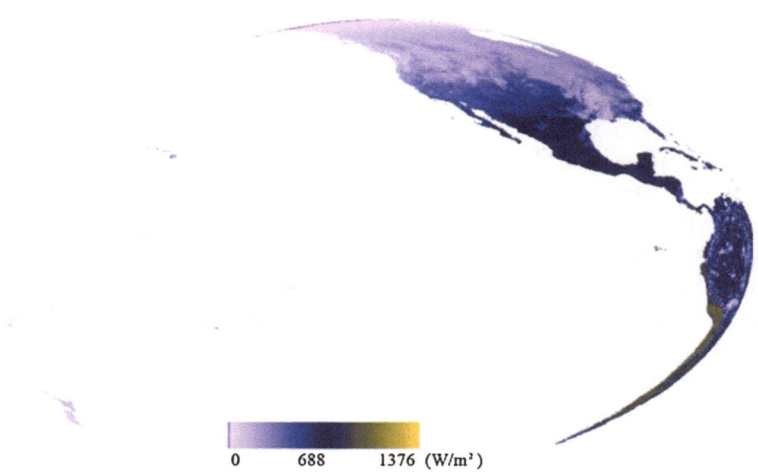

0   688   1376 (W/m^2)

Fig. 5.4 Retrieved ISR using GOES11 data at 18:00 (GMT) on November 12, 2008

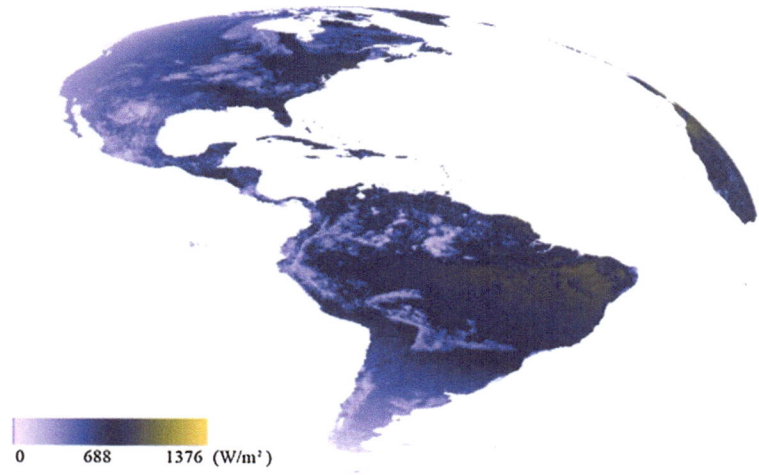

Fig. 5.5 Retrieved ISR using GOES12 data at 14:45 (GMT) on November 12, 2008

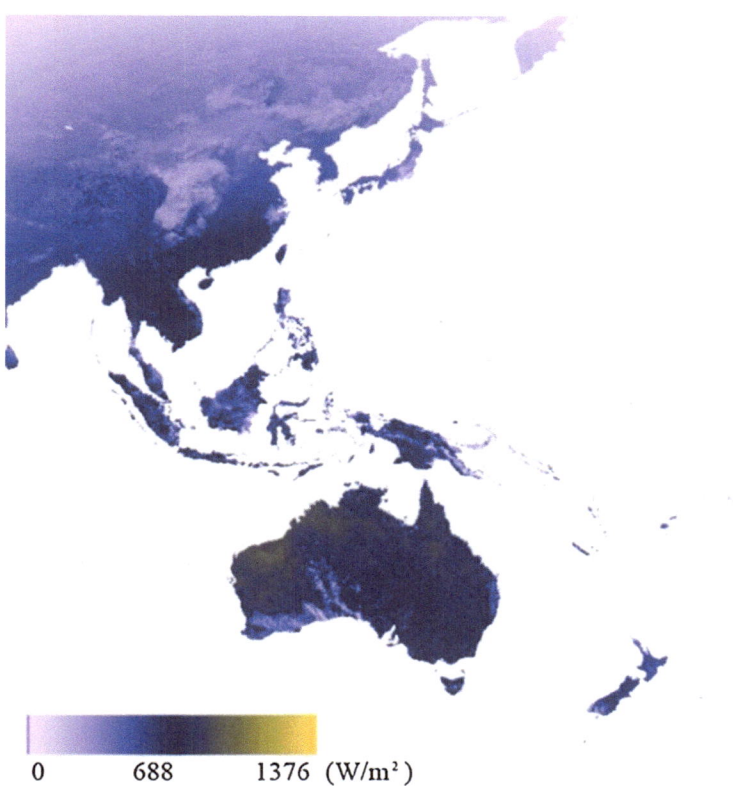

Fig. 5.6 Retrieved ISR using MTSAT data at 03:30 (GMT) on November 12, 2008

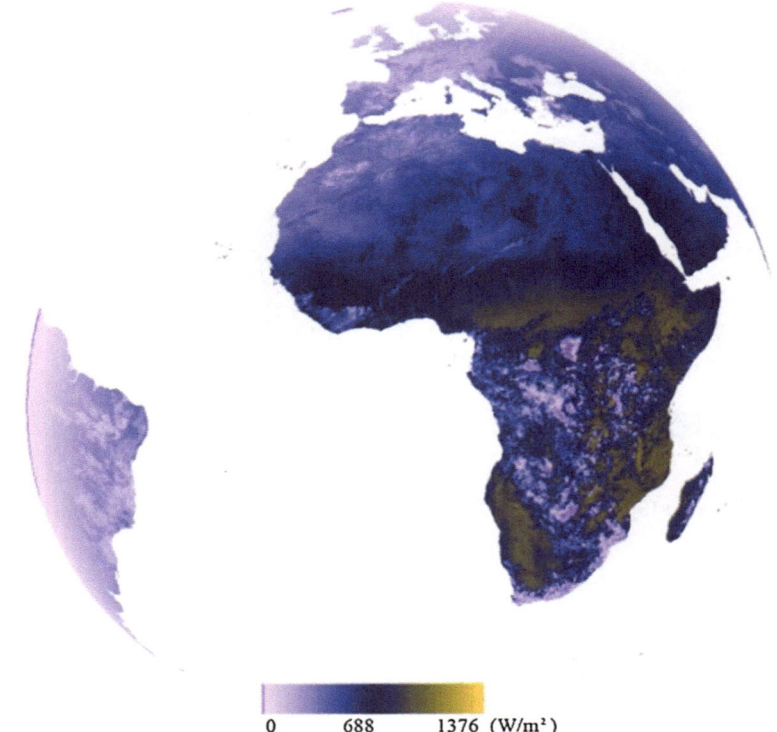

0   688   1376 (W/m²)

Fig. 5.7 Retrieved ISR using MSG2 data at 10:00 (GMT) on November 12, 2008

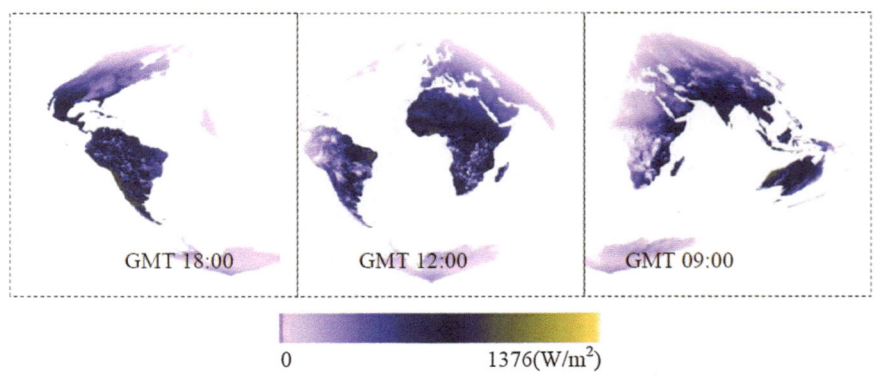

GMT 18:00   GMT 12:00   GMT 09:00

0   1376(W/m²)

Fig. 5.8 The GLASS ISR product using data from multiple polar-orbiting and geostationary satellites on November 12, 2008

Table 5.4 Comparison of retrieved surface insolation from GLASS, international satellite cloud climatology project—flux data (ISCCP-FD), CERES model B, and CALISPO, CERES, clousat, and MODIS (CCCM) enhanced product in 2008

Site	Retrieved DSSR			ISCCP-FD			CERES					
							Model B			CCCM enhanced		
	R2	Bias	RMSE	R2	Bias	RMSE	R2	Bias	RMSE	R2	Bias	RMSE
Bondville	0.87	14.68	104.97	0.71	−7.06	149.88	0.84	12.9	119.5	0.82	−0.5	126.16
Fort Peck	0.84	10.51	102.75	0.69	9.61	150.37	0.81	5.3	112.40	0.80	2.3	115.02
Goodwin creek	0.91	−6.29	99.54	0.64	12.61	184.11	0.69	14.3	172.0	0.66	−3.8	179.35
Penn state	0.85	18.17	109.3	0.7	5.92	152.88	0.87	6.9	107.0	0.86	−8.6	111.18
Sioux falls	0.81	11.52	114.41	0.65	37.83	168.85	0.62	−11.4	167.4	0.58	−37.8	178.77
Boulder	0.81	−12.8	126.38	0.72	6.49	154.96	0.34	−12.0	249.3	0.47	−43.0	214.41
Desert Rock	0.92	−52.4	112.94	0.87	−42.4	125.27	0.52	−24.2	198.0	0.49	−26.6	206.38

for both clear- and cloudy-sky conditions The CALIPSO, CERES, Clousat, and MODIS (CCCM) enhanced products were derived using two-step processes. First, three 333 m resolution CALIPSO profiles of one 1.4 km Cloudsat profile were collocated with each 1 km MODIS imager pixels using geolocation information. Then these 1 km data were collocated with 20 km CERES footprints. It is clear that the GLASS product was more accurate than other two products at these validation sites.

5.5 Summary

High-resolution ISR data are needed for many applications, including ecological and land surface process modeling. However, the majority of existing products have coarse spatial resolutions. The GLASS ISR product has been generated using multiple polar-orbiting and geostationary satellites based on an improved look-up table method with a 5 km spatial resolution and a temporal range from 2008 to 2010. However, this product remains insufficient for long-term radiation budget analyses and other applications. Thus, long-term time series of radiation products must be developed.

Although aerosols and clouds are not input parameters for producing the GLASS ISR product, the uncertainty of the surface reflectance can cause errors, especially for geostationary satellite data. One option is use of the snow-cover product in the algorithm; however, this may result in excessive computation time. In this study, the effect of water vapor absorption on ISR was accounted for using a simple correction function, and the MODIS water vapor column product was used. Missing values or gaps in this product may also affect the precision of the ISR estimates. The influences of cloud types represent an additional source of errors for ISR estimates. Further efforts will be made to address these issues.

References

Anderson GP, Berk A, Acharya PK, Matthew MW, Bernstein LS, James H, Chetwynd J, Dothe H, Adler-Golden SM, Ratkowski AJ, Felde GW, Gardner JA, Hoke ML, Richtsmeier SC, Pukall B, Mello JB, Jeong LS (1999) MODTRAN4: radiative transfer modeling for remote sensing. In: Anton K, John DG (eds) Proceedings of SPIE, pp 2–10
Arking A (1991) The radiative effects of clouds and their impact on climate. Bull Am Meteorol Soc 72:795–953
Chen L, Yan G, Wang T, Ren H, Calbó J, Zhao J, McKenzie R (2012) Estimation of surface shortwave radiation components under all sky conditions: modeling and sensitivity analysis. Remote Sens Environ 123:457–469
Fu Q, Liou KN (1993) Parameterization of the radiative properties of Cirrus clouds. J Atmos Sci 50:2008–2025
Gueymard CA (2003) Direct solar transmittance and irradiance predictions with broadband models. Part I: detailed theoretical performance assessment. Sol Energy 74:355–379

Gui S, Liang S, Wang K, Li L, Zhang X (2010) Assessment of Three Satellite-Estimated Land Surface Downwelling Shortwave Irradiance Data Sets. Geoscience and Remote Sensing Letters, IEEE 77:776–780

Hoch SW, Whiteman CD (2010) Topographic effects on the surface radiation balance in and around Arizona's Meteor Crater. J Appl Meteorol Climatol 49:1114–1128

Huang G, Wang W, Zhang X, Liang S, Liu S, Zhao T, Feng J, Ma Z (2013) Validation of GlASS-DSSR products using surface measurements collected in arid and semi-arid region of China, Int J Digital Earth (in press)

Hubanks PA, King MD, Platnick S, Pincus R (2008) The MODIS Atmosphere L3 Grided Product. In: Algorithm theoretical basis document, No: 30

Jin Z, Yezheng W, Gang Y (2005) General formula for estimation of monthly average daily global solar radiation in China. Energy Convers Manage 46:257–268

Laszlo I, Ciren P, Liu HQ, Kondragunta S, Tarpley JD, Goldberg MD (2008) Remote sensing of aerosol and radiation from geostationary satellites. Adv Space Res 41:1882–1893

Liang S (2004) Quantitative remote sensing of land surface. Wiley, Jew Jersey

Liang S, Li X, Wang J (2012). Advanced Remote Sensing. Terrestrial Information Extraction and Applications. Academic Press

Liang S, Zheng T, Liu R, Fang H, Tsay SC, Running S (2006) Estimation of incident photosynthetically active radiation from moderate resolution imaging spectrometer data. J Geophys Res Atmos 111:D15208

Marshak A, Davis A (2005) 3D radiative transfer in cloudy atmospheres. Springer, Berlin

Mayer B, Hoch SW, Whiteman CD (2010) Validatin9999 g the MYSTIC three-dimensional radiative transfer model with observations from the complex topography of Arizona's Meteor Crater. Atmos Chem Phys 10:8685–8696

Mayer B, Kylling A (2005) Technical note: the libRadtran software package for radiative transfer calculations—description and examples of use. Atmos Chem Phys 5:1855–1877

Otkin JA, Anderson MC, Mecikalski JR, Diak GR (2005) Validation of GOES-based insolation estimates using data from the US climate reference network. J Hydrometeorol 6:460–475

Paulescu M, Schlett Z (2003) A simplified but accurate spectral solar irradiance model. Theoret Appl Climatol 75:203–212

Psiloglou BE, Kambezidis HD (2007) Performance of the meteorological radiation model during the solar eclipse of 29 March 2006. Atmos Chem Phys Discuss 7:6047–6059

Psiloglou BE, Santamouris M, Asimakopoulos DN (2000) Atmospheric broadband model for computation of solar radiation at the earth's surface: application to mediterranean climate. Pure Appl Geophys 157:829–860

Ricchiazzi P, Yang S, Gautier C, Sowle D (1998) SBDART: a research and teaching software tool for plane-parallel radiative transfer in the earth's atmosphere. Bull Am Meteorol Soc 79:2101–2114

Ryu Y, Kang S, Moon S-K, Kim J (2008) Evaluation of land surface radiation balance derived from moderate resolution imaging spectroradiometer (MODIS) over complex terrain and heterogeneous landscape on clear sky days. Elsevier, Amsterdam

Schaaf C, Gao F, Strahler A, Lucht W, Li X, Tsung T, Strugll N, Zhang X, Jin Y, Muller P, Lewis P, Barnsley M, Hobson P, Disney M, Roberts G, Dunderdale M, Doll C, d'Entremont R, Hu B, Liang S, Privette J, Roy D (2002) First operational BRDF, albedo nadir reflectance products from MODIS. Remote Sens Environ 83:135–148

Stamnes K, Tsay SC, Wiscombe W, Jayaweera K (1988) Numerically stable algorithm for discrete-ordinate-method radiative transfer in multiple scattering and emitting layered media. Appl Opt 27:2502–2509

Van Laake PE, Sanchez-Azofeifa GA (2004) Simplified atmospheric radiative transfer modelling for estimating incident PAR using MODIS atmosphere products. Remote Sens Environ 91:98–113

Vermote EF, Tanré D, Deuzé J, Herman M, Morcrette J (1997) Second simulation of the satellite signal in the solar spectrum(6S), 6S user guide version 3

Vermote E, Nazmi, Z, Christopher O (2002) Atmospheric correction of MODIS data in the visible to middle infrared: first results. Remote Sens Environ 83:97–111

Wang D, Liang S (2010) Using multiresolution tree to integrate MODIS and MISR-L3 LAI products. IGARSS 2010:1027–1030

Wild M (2011) Enlightening global dimming and brightening. Bull Am Meteorol Soc 93:27–37

Yang K, He J, Tang W, Qin J, Cheng CCK (2010) On downward shortwave and longwave radiations over high altitude regions: observation and modeling in the Tibetan plateau. Agric For Meteorol 150:38–46

Zhang X, Liang S (2012) Incident solar radiation (Chap 6). In: Liang S, Wang J, Li X (eds) Advanced remote sensing: terestrial information extraction and applications. Elsevier, pp 127–173

Chapter 6
Incident Photosynthetic Active Radiation

Abstract Solar energy in the 400–700 nm spectral range, the so-called photosynthetically active radiation (PAR), plays an important role in photosynthesis, which controls the exchange of water vapor and carbon dioxide between vegetation and the atmosphere. This chapter introduces the GLASS PAR product, covering the algorithm and its validation and analysis. In the first section, a brief introduction will be given. The satellite data used by the GLASS PAR product and detailed algorithm information will be introduced in Sect. 6.2. Quality control and evaluation of the GLASS PAR product using ground measurements will be presented in Sect. 6.3. Preliminary analysis and applications of the GLASS PAR product are shown in Sect. 6.4. A short summary will be given at the end of this chapter.

Keywords PAR · Photosynthetically active radiation · Remote sensing · MODIS · MSG · MTSAT · GOES

6.1 Background

Solar radiation available for photosynthesis, known as photosynthetically active radiation (PAR), constitutes the basic source of energy for biomass by controlling the photosynthetic rate of organisms on land, thus directly affecting plant growth (Frouin and McPherson 2012; Frouin and Murakami 2007; Frouin and Pinker 1995; Liang et al. 2012). The incident PAR, which is defined as the visible part (400–700 nm) of incident shortwave radiation, is also a key parameter in ecological modeling (Running et al. 2004), which controls the exchange of water vapor and carbon dioxide between vegetation and the atmosphere (Liang et al. 2006; Zheng et al. 2008) and makes possible estimation of global oceanic and terrestrial gross primary production (GPP) and net primary production (NPP). Knowing the spatial and temporal distribution of PAR is essential to understand biogeochemical cycles of carbon, nutrients, and oxygen, and to address related

S. Liang et al., *Global LAnd Surface Satellite (GLASS) Products*,
SpringerBriefs in Earth Sciences, DOI: 10.1007/978-3-319-02588-9_6,
© The Author(s) 2014

climate change issues such as evapotranspiration processes and variations (Frouin and McPherson 2012; Frouin and Murakami 2007).

The light use efficiency model is the most common approach to estimating vegetation production. The incident PAR is required by almost all these models such as CASA (Carnegie-Ames-Stanford Approach) (Potter et al. 1993) and the GLO-PEM (Global Production Efficiency Model) (Prince and Goward 1995). The light use efficiency model simplifies photosynthesis through several assumptions: under appropriate environmental conditions (temperature, water, and nutrients), the rate of photosynthesis depends on the absorption of effective solar radiation by the leaves, and the solar energy is changed into chemical energy at a fixed ratio (potential optical energy utilization rate) by plants. Thus, GPP can be calculated using the following equation.

$$GPP = f\text{PAR} \times \text{PAR} \times \varepsilon_{max} \times f \qquad (6.1)$$

where $f\text{PAR}$ is the fraction of absorbed PAR, ε_{max} is the potential light use efficiency and f is the limitation imposed on light use efficiency by various environmental stresses. $f\text{PAR}$ is available as a product from many sensors such as the Moderate Resolution Imaging Spectroradiometer (MODIS), but PAR is not.

In the CASA and GLO-PEM models, PAR is always estimated by multiplying the incident shortwave radiation (ISR) by a conversion factor approximately 0.5. Because most global radiative flux dataset do not include the PAR products, most users have to generate PAR by multiplying the incident surface shortwave radiation by an empirical conversion factor. Frouin and Pinker (1995) concluded that the conversion factor between ISR and PAR varies from 0.45 to 0.5 using the derived global PAR map from ISCCP C1 data with 250 km spatial resolution, in which ISR and PAR are derived separately. Moreover, several studies from ground measurements have also indicated that this conversion factor is not a constant (Dye 2004; Jacovides et al. 2003). Jacovides et al. (2003) found that the conversion factors ranged from 0.460 to 0.501 for the hourly values under cloudy-sky conditions. Wang et al. (2007) investigated the ratio of PAR to ISR using three ground measured data sets and found a dependence of the ratio of PAR to ISR on altitude at both hourly and daily scales. The results suggest that the necessity of considering the altitude dependency of PAR/ISR.

PAR is measured at many observation networks, including the Surface Radiation Budget Network (SURFRAD) (Augustine et al. 2000, 2005) and FLUXNET (Baldocchi et al. 2001). Although ground-based measurement data are available, these data remain insufficient for various applications, for instance carbon cycle modeling and radiation budget assessment. Remote sensing is the most practical way mapping PAR globally because of its large spatial coverage.

Algorithms for PAR estimation from satellite measurements have been widely reported (Eck and Dye 1991; Frouin and McPherson 2012; Frouin and Murakami 2007; Frouin and Pinker 1995; Liang et al. 2006; Pinker and Laszlo 1992a; Su et al. 2007; Van Laake and Sanchez-Azofeifa 2004; Zhang and Liang 2012; Zheng et al. 2008). Eck and Dye (1991) estimated monthly PAR using ultraviolet reflectance instead of radiance in the visible and infrared bands with 500 km spatial resolution from the Total Ozone Mapping Spectrometer (TOMS). The method proposed by Pinker and Laszlo (1992b), which was used to generate the global International Satellite Cloud Climatology Project (ISCCP) PL and GEWEX SRB global shortwave radiation products, was originally developed to estimate shortwave flux (0.2–4.0 μm) by spectral integration of model results in 0.2–4.0, 0.4–0.5, 0.5–0.6, 0.6–0.7, and 0.7–4.0 μm spectral intervals. The main idea of this algorithm is to establish the relationship atmospheric transmissivity and TOA broadband reflectivity by matching the model-derived TOA reflectivity associated with a given atmospheric and surface condition to that observed by the satellite, and then determine the PAR.

Recently, a global, 13-year record of daily averaged PAR at the ocean surface (9 km resolution) has been generated from SeaWiFS, MODIS/Aqua, and MODIS/Terra data. This SeaWiFS- and MODIS-derived PAR product was computed as the difference between the incident solar flux between 400 and 700 nm (known) and the reflected flux (measured), taking into account atmospheric absorption (Frouin and Murakami 2007). However, no corresponding high resolution global PAR over land is available.

The fraction of incident PAR absorbed by vegetation has been provided as one of the land surface variables from MODIS data onboard Terra and Aqua, but not PAR itself. The MODIS science team currently has to disaggregate the coarse-resolution ($1.00° \times 1.25°$) reanalysis solar radiation product of the Global Modeling and Assimilation Office (GMAO) as the forcing data to produce the 1 km PSN/NPP product (MOD17) (Zhao et al. 2006). Thus, a high spatial and temporal resolution PAR product is urgently needed, especially over land. The necessity of a high resolution PAR product is confirmed by significant influences of PAR on GPP/NPP estimation (Hicke 2005).

This chapter will provide a detailed introduction to the GLASS PAR product that has a spatial and temporal resolution of 5 km and 3 h, respectively, including the algorithm and its validation and analysis. The GLASS PAR product was generated from both polar-orbiting and geostationary satellite data. The geostationary satellite data have fine temporal resolution, but limited spatial coverage. On the other hand, the polar-orbiting satellite data offer fine spatial coverage, but have coarse temporal resolution. Thus, the geostationary data can be used to capture the diurnal variations of solar radiation, which are good for calculating the daily integrated solar radiation (Laszlo et al. 2008). The radiation estimation precision will be improved if the polar-orbiting and geostationary satellite data are used simultaneously. The satellite data used in the GLASS PAR product are the same as in the GLASS ISR product and include the MODIS, Meteosat Second Generation (MSG) SEVIRI, Multi-Functional Transport Satellite (MTSAT)-1R, and Geostationary Operational Environmental Satellite (GOES) Imager. The spatial coverage of the geostationary satellite used is shown in Fig. 5.1.

The GLASS PAR product is generated using an improved look-up table approach using the top-of-atmosphere (TOA) radiance of the visible band for each sensor, similarly to the GLASS ISR product. The effects of ozone and the water vapor absorption, surface elevation, and the Bi-directional Reflectance Distribution Function (BRDF) are taken into account in the algorithm. Detailed information on the GLASS PAR product will be provided in the following section. The data sources utilized in the GLASS PAR product and the algorithm itself are given in Sect. 6.2. Product characteristics, quality control, and evaluation are presented in Sect. 6.3. Section 6.4 presents a preliminary analysis and applications of the GLASS PAR product. A summary is provided at the end of this chapter.

6.2 Algorithms

Similarly to the GLASS ISR product, the main idea of the GLASS PAR product is to establish the relationship between the PAR and the TOA radiance through radiative transfer simulation using the Moderate Resolution Transmission (MODTRAN) radiative transfer software (Anderson et al. 1999). In this section, the data sources used in the GLASS PAR product and detailed information on the GLASS PAR algorithm will be presented.

6.2.1 Data Sources

The data sources used by the GLASS PAR product are the same as in the GLASS ISR product, including polar-orbiting satellite data, geostationary satellite data, elevation data, and ground measurement data. The polar-orbiting satellite data come from the MODIS. The approach is to match MODIS observed TOA radiance (band 3) associated with specific atmospheric and surface conditions to the modeled radiance, and then estimate the incident PAR at the surface. Besides the MODIS TOA radiance data, MODIS high-level land surface products are also used for GLASS PAR estimation, including the MODIS geolocation product (MOD03 and MYD03), the land surface reflectance product (MOD09A1) (Vermote et al. 2002), and the BRDF/albedo parameter product (MCD43B1) (Schaaf et al. 2002). The geostationary satellite data used in the GLASS PAR product include the Meteosat Second Generation (MSG) SEVIRI, the MTSAT-1R, and the GOES Imager from both GOES-East and GOES-West data. The visible bands of these geostationary satellite data are used for GLASS PAR estimation. Moreover, the elevation data from GTOPO30 are used to reduce the influence of topography on PAR estimation. The ground measured PAR data are used to evaluate the GLASS PAR product.

6.2.2 Algorithm Description

6.2.2.1 Estimating PAR Based on the Look-up Table Method

Radiative transfer codes can be used to estimate incident broadband solar radiation or PAR if a complete set of retrieved cloud and atmospheric parameters from other sources is available. This approach has been used for estimating PAR from the CERES (Su et al. 2007; Wielicki et al. 1998) and the GEWEX (Pinker et al. 2003). This approach has a clear physical basis, but use of multiple atmospheric and surface products does not make it possible to generate the high spatial resolutions needed to meet the requirements of land applications. The uncertainties of these atmospheric properties can also affect the accuracy of the incident solar radiation products.

The alternative approach is to establish the relationship between the TOA radiance and surface incident solar radiation based on extensive radiative transfer simulations. To improve computational efficiency, a look-up table (LUT) is used to determine incident solar radiation at the surface based on TOA observations. Liang et al. (2006) presented a LUT method to estimate PAR under different illumination geometries and atmospheric conditions based on MODTRAN4 simulations. In follow-up studies, a series of refinements and improvements were made. For example, the MODIS surface reflectance product (MOD09) was used to map PAR over China from MODIS data (Liu et al. 2008). Zheng et al. (2008) use this algorithm to estimate PAR from GOES data taking into account topographic effects. Zhang et al. (2013) further extended this algorithm to estimate GLASS global land surface PAR from multiple satellite data. The main procedure of the GLASS PAR product is shown in Fig. 6.1. Water vapor absorbs infrared radiation, and the influence of water vapor on the estimation of the incident PAR is small and can be ignored. Thus, the major difference between the GLASS ISR and PAR estimation algorithms is that the GLASS ISR is corrected using the MODIS water vapor product (MOD08).

MODTRAN4 (Anderson et al. 1999) is used to simulate radiative transfer of the solar spectrum to generate the GLASS PAR product. The parameters used in MODTRAN are summarized in Table 5.1. Radiative transfer is simulated with a selected aerosol model and cloud type at different view geometries. The downward spectral flux at the surface resulting from the MODTRAN4 simulation can be integrated from 400 to 700 nm to estimate PAR. The upward top-of-atmosphere (TOA) spectral radiance for each sensor can be integrated with the sensor's channel response function to calculate the channel radiance. The workflow for the LUTs is as follows:

First, the TOA reflectance and radiance for different sensors are calculated based on the DN numbers using the calibration coefficients. Second, the surface reflectance is estimated for the different sensors or the satellite-derived surface reflectance products are directly used. Third, the TOA radiance is estimated for each atmospheric condition from the clearest to the cloudiest (high cloud extinction

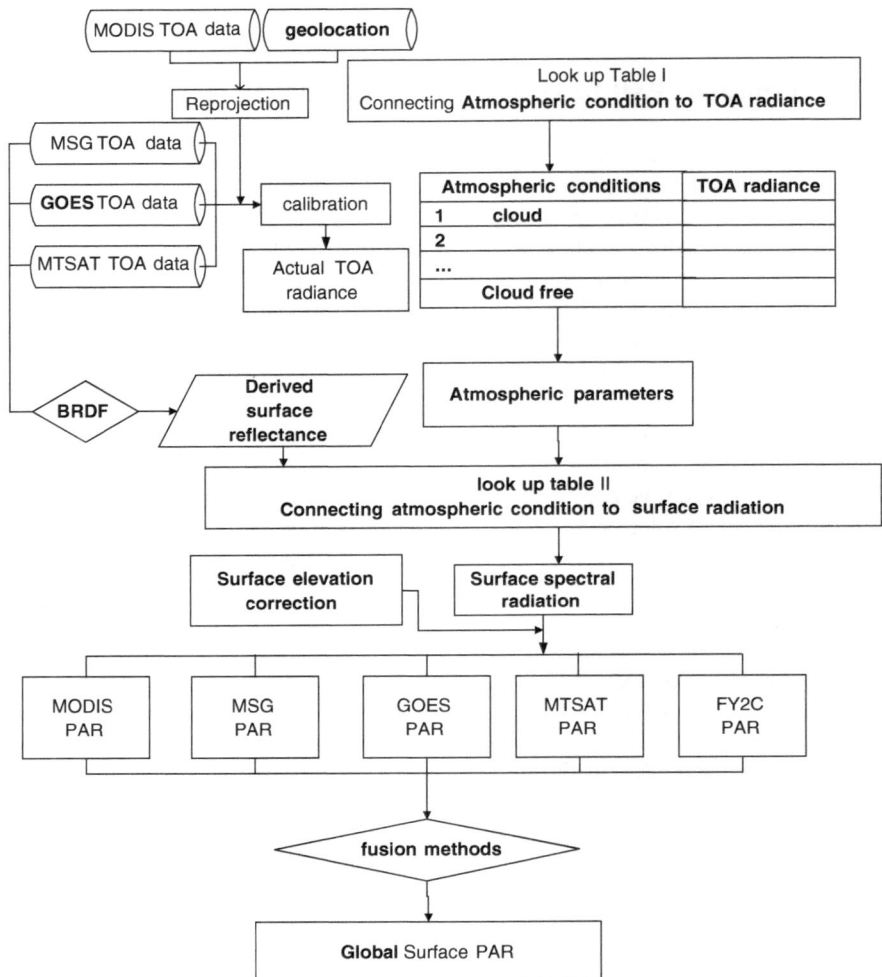

Fig. 6.1 Flow diagram of GLASS PAR estimation

coefficient). Then the actual TOA radiance calculated from the sensors is compared to the series of simulated radiance values for different atmospheric conditions to retrieve the atmospheric index. Finally, with the estimated atmospheric condition index and surface reflectance, the spectral flux can be estimated and integrated to yield PAR. The details of creating the LUT and determining surface reflectance are introduced in Sect. 5.2.2.

Similarly to the GLASS ISR product, the MODIS data are used to derive the incident PAR at the surfaces at north and south latitudes greater than 60°, while at lower latitudes PAR is calculated through a combination of the polar-orbiting and geostationary satellite-derived radiation products. Please refer to Sect. 5.2.2.5 for details of the combination procedure.

6.2.2.2 Sensitivity Analysis

Solar radiation is attenuated as it passes through the atmosphere. Water vapor and carbon dioxide absorb infrared radiation, ultraviolet radiation is absorbed by ozone, and the shorter range of the visible spectrum is scattered by aerosols and molecules. To perform a better assessment of the interactions of these atmospheric input variables in the LUT, a sensitivity analysis of the atmospheric profiles, ozone level, water vapor amount, surface elevation, and aerosol model are carried out. The six standard atmosphere profiles (tropical, mid-latitude winter, mid-latitude summer, sub-arctic winter, sub-arctic summer, and 1976 US standard) were employed with specified atmospheric variables. The influences of the atmospheric profiles on PAR estimation are illustrated in Fig. 6.2. As shown in Fig. 6.2, the influence of atmospheric profiles is small, with the highest value for the arctic winter profiles. Thus, it can be concluded that PAR estimation using the LUT method is not sensitive to atmosphere profiles for this algorithm, which is consistent with the results obtained by Liang et al. (2006).

The sensitivity experiments used to determine the dependence of PAR on ozone levels are shown in Fig. 6.3 for mid-latitude summer atmospheric profiles using a

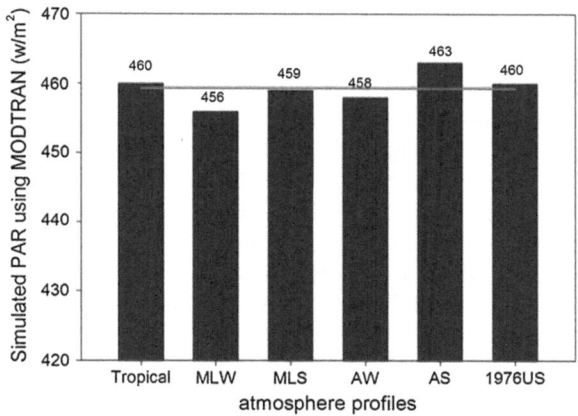

Fig. 6.2 Sensitivity of PAR estimates to atmospheric profiles. The *dark yellow line* represents the average values of simulated PAR using different atmospheric profiles. The atmospheric condition is represented by a mid-latitude winter atmosphere with a rural aerosol type. Other input atmospheric variables are set as constants (solar zenith angle, 0°; visibility, 90 km; surface albedo, 0.2; cloud extinction coefficient, 1 km^{-1}). Trp, MLW, MLS, AW, AS, and 1976 US represent the tropical, mid-latitude winter, mid-latitude summer, arctic winter, arctic summer, and 1976 US standard atmosphere profiles, respectively

Fig. 6.3 Sensitivity of PAR to variations in the amounts of ozone and water vapor. The atmospheric condition is represented by a mid-latitude summer atmosphere with a rural aerosol type. Other input atmospheric variables are set as constants (water vapor amount, 2.0 g/m^2; visibility, 90 km; albedo, 0.20)

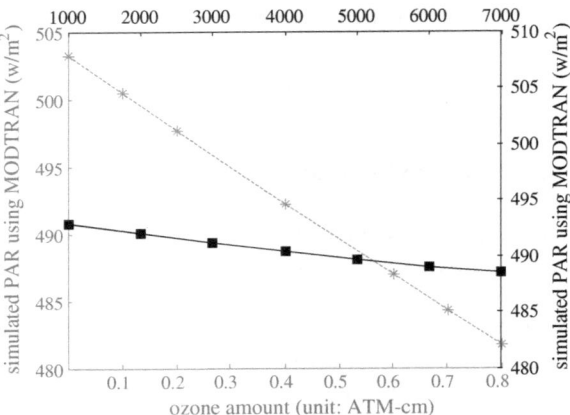

rural aerosol model under cloud-free conditions and a visibility of 90 km, which is considered a very clear atmospheric condition. The other atmospheric parameters are kept constant in MODTRAN. As shown in Fig. 6.3, the PAR estimates vary by approximately 20 Wm^{-2}, at the surface with increasing ozone from 0.0 to 0.8 atm-cm. However, the amount of ozone does not change significantly under most conditions. Under other relatively hazy or cloudy air conditions, the influences of the ozone level will be less. Thus, ozone is kept a constant in the simulation, and the effects of ozone are not corrected for in the proposed algorithm. The effects of water vapor on PAR estimation are also shown in Fig. 6.3. It is readily apparent that the influence of water vapor on PAR estimation is relatively small and can be ignored.

The effects of surface elevation on PAR are also important and cannot be ignored, and numerous reports on this topic have been published (Wang et al. 2005). Surface elevation effects are considered as a dimension in the LUT in the proposed algorithm. The variations in the PAR estimates with changes in surface elevation are described in Fig. 6.4. The conclusion is that the influence of surface altitude on PAR is significant under both cloudy and cloud-free conditions and cannot be neglected. Surface elevation may cause an error of up to 100 Wm^{-2} and is a significant source of error in PAR estimation. For example, this effect is the most likely reason that a number of current radiation products have large errors over the Tibetan Plateau.

To assess the effects of aerosol models on PAR simulations using MODTRAN, four aerosol models were selected to represent atmospheric turbidity in MOD-TRAN. The sensitivities of PAR to selected aerosol models (rural, maritime, urban, and tropospheric modes) under cloud-free conditions with 23 km visibility are shown in Fig. 6.5. It can be concluded that the rural, maritime, and tropospheric aerosol models do not have a significant influence on PAR, whereas the urban aerosol model does have an effect. The urban aerosol models show the largest difference, reaching approximately 29 Wm^{-2} for PAR due to the strong

Fig. 6.4 Sensitivity of PAR at the surface to surface altitude under different atmospheric conditions (100 km visibility, 5 km visibility, 1 km^{-1} cloud extinction, and 7 km^{-1} cloud extinction of mid-latitude summer atmosphere with a rural aerosol type)

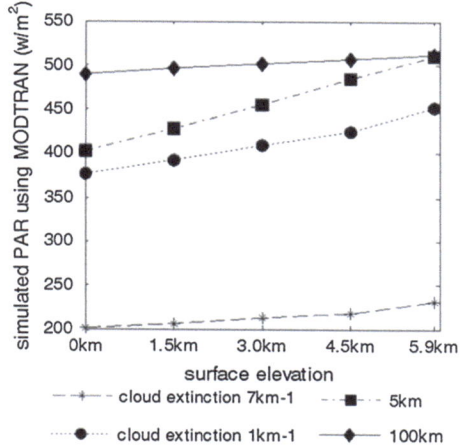

Fig. 6.5 Sensitivity of PAR to aerosol modes under different solar zenith angles. The atmospheric condition is represented by a mid-latitude summer atmosphere. Other input atmospheric variables are kept constant (water vapor, 2.0 gm^{-2}; visibility, 23 km; albedo, 0.20)

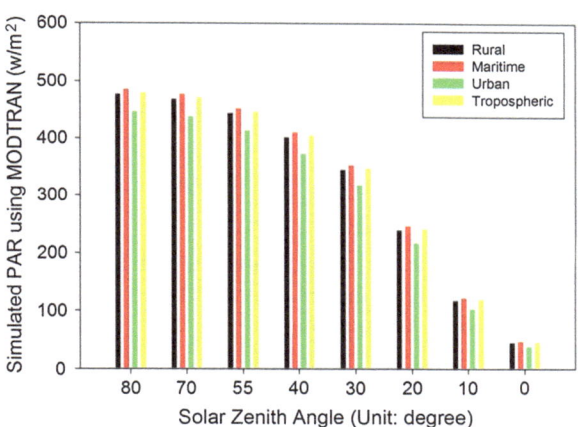

absorption of solar radiation by urban aerosols. The PAR retrieval will be affected by the distinct characteristics and size distributions of different aerosol models under cloud-free conditions. However, the influences of the aerosol models on the estimation of solar radiation at the surface are not implemented in this proposed algorithm because the aerosol influences are insignificant under cloudy conditions and significant under clear-sky conditions.

6.2.2.3 Integrated Daily PAR

The integrated daily PAR is often used in estimation of ecosystem productivity such as GPP/NPP estimation. However, satellite imagery is simply a snapshot of the Earth, and, consequentially, the direct estimation of solar radiation from the imagery is usually the instantaneous value of that snapshot. For geostationary satellites with frequent revisit times (e.g., every 15 min, 60 min, 3 h for MSG2, MTSAT, and GOES full dish imagery respectively), temporal scaling from instantaneous values to daily values is straightforward. Nevertheless, estimating the daily mean solar radiation from the few instantaneous values retrieved from polar-orbiting satellite sensors is a very complex undertaking (e.g., for MODIS). MODIS data are used to derive the GLASS PAR product for surfaces with north and south latitudes greater than 60°. Thus, it is important to reduce errors to calculate the daily integrated PAR from limited observations.

One solution is to interpolate the atmospheric parameters first. For example, Zhang et al. developed a method to derive the daily mean PAR from interpolated inputs of atmospheric and surface parameters (Zhang et al. 1995). It is also possible to calculate daily value directly from instantaneous values (Liang et al. 2006). Zheng and Liang (2011) developed a Bayesian Method of Ration (MOR) method to combine sparse instantaneous PAR observations with a prior knowledge to predict daily values. The GLASS daily PAR product is estimated using the algorithm proposed by Wang et al. (2010). The main operations involved in this method can be described as follows:

Given one instantaneous PAR retrieval InstPAR(T_{overpass}) at time T_{overpass}, the instantaneous PAR retrievalInstPAR(t) at any time of day t can be interpolated as:

$$\text{InstPAR}(t) = \text{InstPAR}(T_{\text{overpass}}) \frac{\sin \frac{(t-T_{\text{sunrise}})\pi}{T_{\text{sunset}}-T_{\text{sunrise}}}}{\sin \frac{(T_{\text{overpass}}-T_{\text{sunrise}})\pi}{T_{\text{sunset}}-T_{\text{sunrise}}}} \tag{6.2}$$

where T_{sunrise} is the time of sunrise, and T_{sunset} is the time of sunset.

After obtaining the PAR at any time, the daily mean PAR can be easily calculated using the equation:

$$\text{DailyPAR} = \frac{1}{T} \int_{T_{\text{sunrise}}}^{T_{\text{sunset}}} \text{InstPAR}(t)dt \tag{6.3}$$

Wang et al. (2010) also designed an interpolation method to handle cases with more than one instantaneous PAR value available. Consider an example of two observations within one day at T_1 and T_2 respectively. From sunrise to T_1, PAR takes on the value of InstPAR$_{T_1}(t)$, and from T_2 to sunset, PAR has the value of InstPAR$_{T_2}(t)$. Between T_1 and T_2, the PAR value is calculated by the weighted average of two sinusoidally interpolated values:

$$\text{InstPAR}(t) = \frac{T_2 - t}{T_2 - T_1} \text{InstPAR}_{T_1}(t) + \frac{t - T_1}{T_2 - T_1} \text{InstPAR}_{T_2}(t) \tag{6.4}$$

Similarly, this approach can be extended to cases with N observations at $T_1, \ldots, T_N$. By substituting Eq. (6.3) into Eq. (6.4), the daily mean PAR can be calculated as:

$$\text{DailyPAR} = \frac{1}{T} \left[\int_{T_{\text{sunrise}}}^{T_1} \text{InstPAR}_{T_1}(t) \, dt \right.$$

$$+ \sum_{i=1}^{N-1} \int_{T_i}^{T_{i+1}} \left(\frac{T_{i+1} - t}{T_{i+1} - T_i} \text{InstPAR}_{T_i}(t) + \frac{t - T_i}{T_{i+1} - T_i} \text{InstPAR}_{T_{i+1}}(t) \right) dt$$

$$\left. + \int_{T_N}^{T_{\text{sunset}}} \text{InstPAR}_{T_N}(t) \, dt \right]$$

(6.5)

6.3 Product Generation, Quality Control and Validation

6.3.1 Product Generation

The GLASS PAR product was generated using the algorithms and procedures described above using the GLASS production system (Zhao et al. 2013). The global GLASS PAR product was generated from the estimated PAR of the selected polar-orbiting and geostationary satellites. Thus, GLASS PAR product includes not only the global fused product but also that for each satellite. The spatial and temporal resolutions are summarized in Table 5.2.

6.3.2 Quality Control

Similarly to the GLASS ISR product, the quality of the GLASS PAR product was primarily dependent on atmospheric conditions, snow cover, and surface reflectance. The quality of the GLASS PAR product is represented by a scientific dataset, which includes information on cloud state, cloud shadow, surface reflectance quality, and snow cover within a 16-bit unsigned integer. A detailed description of GLASS PAR quality control information is given in Table 5.3.

6.3.3 Validation

The GLASS instantaneous PAR product was validated using the ground measurements data from SURFRAD, Ameriflux, and AERONET. Detailed information on the sites and validation results are summarized in Table 6.1. The overall R-squared value at these sites was 0.84. The bias and RMSE of the GLASS PAR were 5.3 and 49 Wm^{-2}, respectively.

Table 6.1 Validation results of retrieved PAR at the selected sites

Site	Radiation network	PAR		
		R^2	BIAS	RMSE
Bondville	SURFRAD	0.86	4.6	45
FortPeck	SURFRAD	0.82	1.6	46
Goodwin creek	SURFRAD	0.91	4.2	38
Penn state	SURFRAD	0.86	9.4	44
Sioux falls	SURFRAD	0.86	2.4	43
Boulder	SURFRAD	0.78	−7.6	58
Desert rock	SURFRAD	0.89	−30	51
ARM-SGP main	AmeriFlux	0.88	16	45
Audubon research	AmeriFlux	0.87	24	56
Brookings	AmeriFlux	0.84	33	55
FIR	AmeriFlux	0.78	2	61
Flagstaff managed forest	AmeriFlux	0.77	−19	68
Flagstaff unmanaged Forest	AmeriFlux	0.88	−4	44
Neustift	CarbonEuropeIP	0.83	−5	48
Lonzee	CarbonEuropeIP	0.74	9	48
Vielsalm	CarbonEuropeIP	0.79	22	47
Laegern	CarbonEuropeIP	0.83	−4	49
Oensingen crop	CarbonEuropeIP	0.86	−9	47
Bily Kriz forest	CarbonEuropeIP	0.83	24	48
Bily Kriz grassland	CarbonEuropeIP	0.82	25	49
CZECHWET	CarbonEuropeIP	0.86	11	41
Moldova	AERONET	0.85	7	47

6.4 Preliminary Analysis and Applications

6.4.1 Spatial and Temporal Changes in GLASS PAR

Surface solar radiation (including ISR and PAR) show certain trends at the global scale in terms of spatial and temporal variations. The total yearly surface solar radiation absorbed by land surfaces is always greater in low-latitude areas than in high-latitude areas, while the seasonal variation in high-latitude areas is more significant than in low-latitude areas. Theoretically, on the vernal and autumnal equinoxes, the Equator is exposed to direct sunlight and thus receives the most radiation on the planet, while the radiation level progressively decreases toward the Polar regions. Considering clouds and other factors, this trend might differ under real conditions (Zhang and Liang 2012).

In summer, daytime in the Northern Hemisphere is long and increases in regions at higher latitudes. The Arctic Circle even has a Polar day. High-latitude regions receive lower solar radiation due to the effect of the solar elevation angle, but the total amount of solar radiation that they receive is large as a result of longer sunshine duration. In winter, the daytime length in the Southern Hemisphere

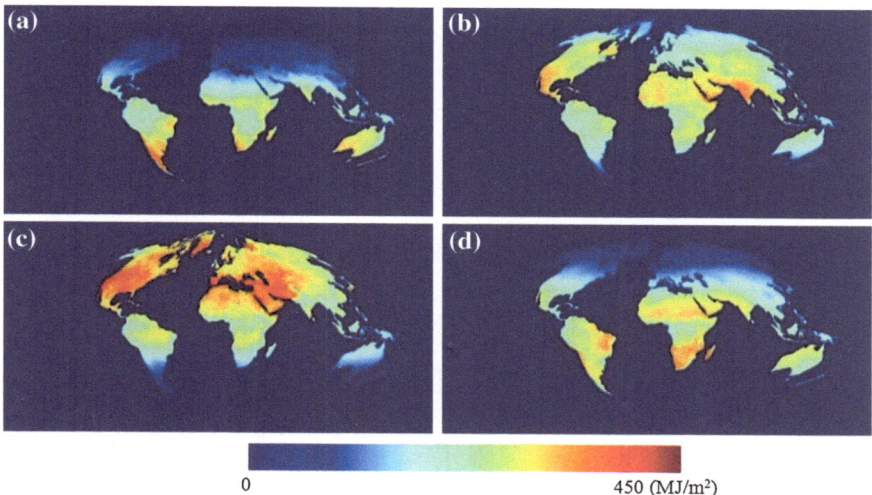

Fig. 6.6 Global land surface monthly integrated incident PAR in 2008 at 5 km spatial resolution calculated from the GLASS PAR product **a** January, **b** April, **c** July, and **d** October

increases, and higher latitudes in the Southern Hemisphere gives rise to longer daytimes. Polar days occur south of the Antarctic Circle. During this period, the radiation received by the Antarctic Circle and high-latitude areas reaches its maximum. Figure 6.6 shows the pattern of the temporal and spatial distributions of monthly integrated PAR in January, April, July, and October 2008 based on the GLASS PAR product. The temporal variations of monthly integrated PAR for different regions, including global, Asia, North America, Europe, South America, Australia, and Africa, are presented in Fig. 6.7.

6.4.2 GPP Estimation Using GLASS PAR Over China

The vegetation gross carbon uptake (GPP) plays an important role in quantifying the global carbon cycle. PAR is an essential variable for GPP estimation as discussed in the first section. The EC-LUE (Eddy Covariance-Light Use Efficiency) model is a light use efficiency model based on carbon flux data measured at eddy covariance stations (Yuan et al. 2007, 2010). Eleven eddy covariance sites located in northern China were used to validate the accuracy of the GPP simulated by the EC-LUE model across different vegetation types of cropland, grass, and broadleaf forest. Eddy covariance systems directly measure net ecosystem exchange (NEE) rather than GPP, thus half hourly NEE, air temperature, and friction velocity must be processed to obtain daily GPP using the data analyses procedures presented by Yuan et al. (2010).

Fig. 6.7 Temporal variation of land surface monthly integrated PAR in different regions based on the GLASS PAR product

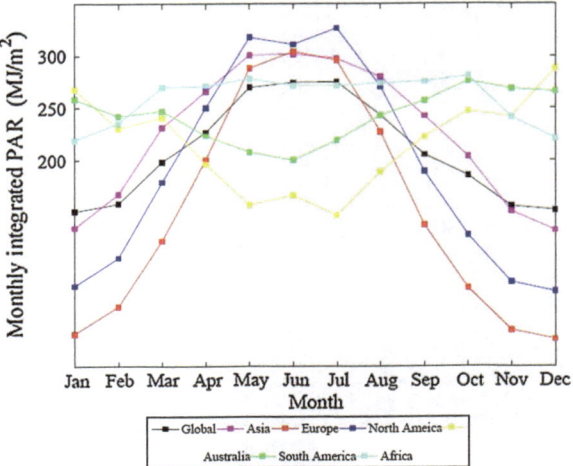

The spatial pattern of GPP was explored using the EC-LUE model driven by GLASS PAR, normalized difference vegetation index (NDVI), air temperature, evaporation, and net radiation. Estimation of Chinese terrestrial GPP in 2008 and 2009 are on average 5.5 PgC/year. Figure 6.8 shows that GPP driven by GLASS PAR tends to overestimate EC GPP especially at lower values, but the regression line is closest to the 1:1 line with high correlation (0.6). The spatial pattern of EC-LUE GPP (Fig. 6.9) demonstrates that the major distribution trend of Chinese terrestrial gross carbon uptake gradually decreases from the southeast coast to the northwest with broadleaf forests in South China exhibiting the highest GPP, as much as 1809 g C/m^2/year. In contrast, arid regions in northwest China have barely any productivity due to cold temperatures in winter, water stress, sparse vegetation cover, which constrain photosynthesis.

Fig. 6.8 Comparison of EC-LUE GPP driven by GLASS PAR with measured GPP at eddy covariance towers

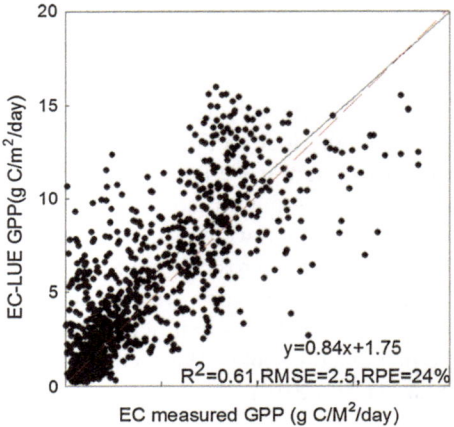

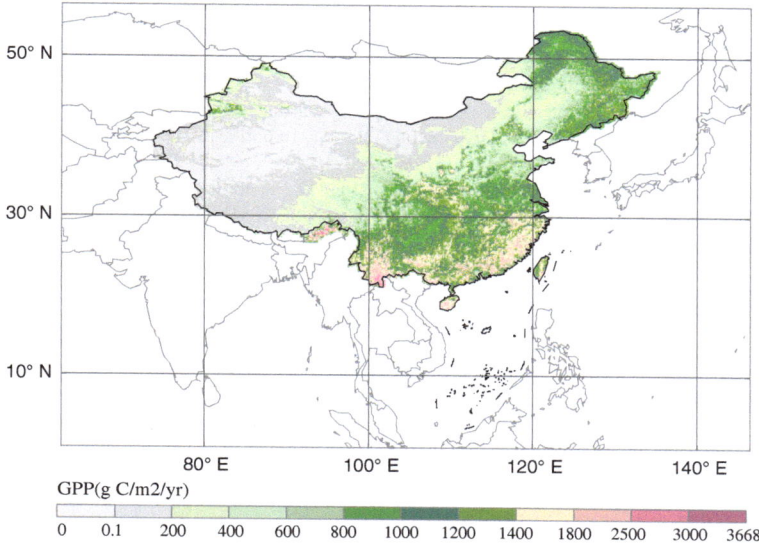

GPP(g C/m2/yr)

| 0 | 0.1 | 200 | 400 | 600 | 800 | 1000 | 1200 | 1400 | 1800 | 2500 | 3000 | 3668 |

Fig. 6.9 Spatial pattern of the annual averaged 2 year (2008–2009) EC-LUE GPP

6.5 Summary

The GLASS PAR product has been generated from multiple polar-orbiting and geostationary satellite data based on an improved look-up table method with a 5 km spatial resolution and a temporal coverage from 2008 to 2010. Given geographically varying biases, however, the limited site evaluation performed does not necessarily represent the overall quality of the GLASS PAR product.

Further refinement of data quality, including both satellite and ground measurements, and individual PAR estimation models should lead to closer agreement and better accuracy of the GLASS PAR product. Such efforts, however, are hindered by a current lack of high precision satellite products and surface measurements. Therefore, the establishment of a long-term, global network of ground-based PAR measurements is needed.

The diffuse component of PAR is as essential as the total PAR, because a higher diffuse portion of PAR is associated with a higher rate of forest net ecosystem exchange of CO_2. Currently, none of the satellite-derived PAR products provides separate direct and diffuse PAR components except GEWEX SRB data. However, the spatial resolution of the GEWEX PAR product is one degree, which is too coarse for many land applications. Therefore, both direct and diffuse components of the GLASS PAR product should be produced in the future.

References

Anderson GP, Berk A, Acharya PK, Matthew MW, Bernstein LS, James H. Chetwynd J, Dothe H, Adler-Golden SM, Ratkowski AJ, Felde GW, Gardner JA, Hoke ML, Richtsmeier SC, Pukall B, Mello JB, Jeong LS (1999) MODTRAN4: radiative transfer modeling for remote sensing. In Anton K, John DG (eds.) SPIE, pp 2–10

Augustine JA, DeLuisi JJ, Long CN (2000) SURFRAD—A national surface radiation budget network for atmospheric research. Bull Am Meteorol Soc 81:2341–2357

Augustine JA, Hodges GB, Cornwall CR, Michalsky JJ, Medina CI (2005) An update on SURFRAD—the GCOS surface radiation budget network for the continental united states. J Atmos Oceanic Technol 22:1460–1472

Baldocchi D, Falge E, Gu L, Olson R, Hollinger D, Running S, Anthoni P, Bernhofer C, Davis K, Evans R (2001) FLUXNET: A new tool to study the temporal and spatial variability of ecosystem-scale carbon dioxide, water vapor, and energy flux densities. Bull Am Meteorol Soc 82:2415–2434

Dye DG (2004) Spectral composition and quanta-to-energy ratio of diffuse photosynthetically active radiation under diverse cloud conditions. J Geophys Res Atmos 109:D10203

Eck TF, Dye DG (1991) Satellite estimation of incident photosynthetically active radiation using ultraviolet reflectance. Remote Sens Environ 38:135–146

Frouin R, McPherson J (2012) Estimating photosynthetically available radiation at the ocean surface from GOCI data. Ocean Sci J 47:313–321

Frouin R, Murakami H (2007) Estimating photosynthetically available radiation at the ocean surface from ADEOS-II global imager data. J Oceanogr 63:493–503

Frouin R, Pinker RT (1995) Estimating photosynthetically active radiation (PAR) at the earth's surface from satellite observations. Remote Sens Environ 51:98–107

Hicke JA (2005). NCEP and GISS solar radiation data sets available for ecosystem modeling: description, differences, and impacts on net primary production. Global Biogeochem Cycles 19

Jacovides CP, Tymvios FS, Asimakopoulos DN, Theofilou KM, Pashiardes S (2003) Global photosynthetically active radiation and its relationship with global solar radiation in the Eastern Mediterranean basin. Theoret Appl Climatol 74:227–233

Laszlo I, Ciren P, Liu HQ, Kondragunta S, Tarpley JD, Goldberg MD (2008) Remote sensing of aerosol and radiation from geostationary satellites. Adv Space Res 41:1882–1893

Liang S, Zheng T, Liu R, Fang H, Tsay SC, Running S (2006) Estimation of incident photosynthetically active radiation from moderate resolution imaging spectrometer data. J Geophys Res Atmos 111:D15208

Liang S, Li X, Wang J (ed) (2012) Advanced remote sensing: terrestrial information extraction and applications. Academic Press, Oxford

Liu R, Liang S, He H, Liu J, Zheng T (2008) Mapping incident photosynthetically active radiation from MODIS data over China. Remote Sens Environ 112:998–1009

Pinker RT, Laszlo I (1992a) Global distribution of photosynthetically active radiation as observed from satellites. J Clim 5:56–65

Pinker RT, Laszlo I (1992b) Modeling surface solar irradiance for satellite applications on a global scale. J Appl Meteorol 31:194–211

Pinker RT, Tarpley JD, Laszlo I, Mitchell KE, Houser PR, Wood EF, Schaake JC, Robock A, Lohmann D, Cosgrove BA, Sheffield J, Duan Q, Luo L, Higgins RW (2003) Surface radiation budgets in support of the GEWEX continental-scale international project (GCIP) and the Gewex Americas prediction project (GAPP), including the North American land data assimilation system (NLDAS) project. J Geophys Res Atmos 108:8844

Potter CS, Randerson JT, Field CB, Matson PA, Vitousek PM, Mooney HA, Klooster SA (1993) Terrestrial ecosystem production: a process model based on global satellite and surface data. Global Biogeochem Cycles 7:811–841

Prince SD, Goward SN (1995) Global primary production: a remote sensing approach. J Biogeogr 22:815–835

Running SW, Nemani RR, Heinsch FA, Zhao M, Reeves M, Hashimoto H (2004). A continuous satellite-derived measure of global terrestrial primary production. BioScience 54

Schaaf C, Gao F, Strahler A, Lucht W, Li X, Tsung T, Strugll N, Zhang X, Jin Y, Muller P, Lewis P, Barnsley M, Hobson P, Disney M, Roberts G, Dunderdale M, Doll C, d'Entremont R, Hu B, Liang S, Privette J, Roy D (2002) First operational BRDF, albedo nadir reflectance products from MODIS. Remote Sens Environ 83:135–148

Su W, Charlock TP, Rose FG, Rutan D (2007) Photosynthetically active radiation from clouds and the earth's radiant energy system (CERES) products. J Geophys Res Biogeosciences 112:G02022

Van Laake PE, Sanchez-Azofeifa GA (2004) Simplified atmospheric radiative transfer modelling for estimating incident PAR using MODIS atmosphere products. Remote Sens Environ 91:98–113

Vermote E, Nazmi Z, Christopher O (2002) Atmospheric correction of MODIS data in the visible to middle infrared: first results. Remote Sens Environ 83:97–111

Wang D, Liang S (2010) Using multiresolution tree to integrate MODIS and MISR-L3 LAI products. IGARSS 2010:1027–1030

Wang K, Zhou X, Liu J, Sparrow M (2005) Estimating surface solar radiation over complex terrain using moderate-resolution satellite sensor data. Int J Remote Sens 26:47–58

Wang Q, Kakubari Y, Kubota M, Tenhunen J (2007) Variation on PAR to global solar radiation ratio along altitude gradient in Naeba Mountain. Theoret Appl Climatol 87:239–253

Wielicki BA, Barkstrom BR, Baum BA, Charlock TP, Green RN, Kratz DP, Lee RB, Minnis P, Smith GL, Takmeng W, Young DF, Cess RD, Coakley JA, Crommelynck DAH, Donner L, Kandel R, King MD, Miller AJ, Ramanathan V, Randall DA, Stowe LL, Welch RM (1998) Clouds and the earth's radiant energy system (CERES): algorithm overview. IEEE Trans Geosci Remote Sens 36:1127–1141

Yuan W, Liu S, Zhou G, Tieszen LL, Baldocchi D, Bernhofer C, Gholz H, Goldstein AH, Goulden ML, Hollinger DY, Hu Y, Law BE, Stoy PC, Vesala T, Wofsy SC (2007) Deriving a light use efficiency model from eddy covariance flux data for predicting daily gross primary production across biomes. Agric For Meteorol 143:189–207

Yuan W, Liu S, Yu G, Bonnefond JM, Chen J, Davis K, Desai AR, Goldstein AH, Gianelle D, Rossi F (2010) Global estimates of evapotranspiration and gross primary production based on MODIS and global meteorology data. Remote Sens Environ 114:1416–1431

Zhang X, Liang S (2012). Incident solar radiation. In Liang S, Wang J, Li X (eds) Advanced remote sensing: terrestrial information extraction and applications, Chapter 6. Elsevier, Amsterdam, pp 127–173

Zhang YC, Rossow WB, Lacis AA (1995) Calculation of surface and top of atmosphere radiative fluxes from physical quantities based on ISCCP data sets: 1. method and sensitivity to input data uncertainties. J Geophys Res Atmos 100:1149–1165

Zhang X, Liang S, Zhou G, Wu H (2013). Mapping global incident downward shortwave radiation and photosynthetically active radiation over land surfaces using multiple satellite data (Submitted). J Geophys Res Atmos

Zhao M, Running SW, Nemani RR (2006) Sensitivity of moderate resolution imaging spectroradiometer (MODIS) terrestrial primary production to the accuracy of meteorological reanalyses. J Geophys Res Biogeosciences 111:G01002

Zhao X, Liang S, Liu S, Yuan W, Xiao Z, Liu Q, Cheng J, Zhang X, Tang H, Zhang X, Liu Q, Zhou G, Xu S, Yu K (2013) The global land surface satellite (GLASS) remote sensing data processing system and products. Remote Sens 5:2436–2450

Zheng T, Liang SL (2011) A Bayesian approach to integrate satellite-estimated instantaneous photosynthetically active radiation product for daily value calculation. J Geophys Res Atmos 116

Zheng T, Liang S, Wang K (2008) Estimation of incident photosynthetically active radiation from GOES visible imagery. J Appl Meteorol Climatol 47:853–868

Chapter 7
Challenges and Prospects

A high-resolution long-term GLASS dataset, including shortwave albedo, longwave emissivity, LAI, incident shortwave radiation, and incident PAR, has been developed and is being distributed to the international research community for evaluation and utilization. This book has documented algorithm development, product characteristics, quality control information, and preliminary results of validation, analysis, and applications.

Many land products have been produced from various satellite data. Because most of these products are based on each individual satellite missions with limited life spans, few land products so far have covered the extended periods of time which are suitable for long-term global and regional environmental change studies. The U.S. National Research Council (NRC) has called for production of the climate data record (CDR) from multiple satellite datasets. A CDR is defined as a time series of measurements of sufficient length, consistency, and continuity to determine climate variability and change. Efforts are currently underway to generate more land products from the GLASS product production system. Integration of multiple satellite data for producing high-level land surface products offers great promise, yet also presents great challenges.

Almost all land products are generated primarily from one instrument algorithm for each product, but as a rule it is almost impossible to identify the best algorithm as most of them perform optimally only under certain conditions. Thus, the accuracy of a specific product is not consistent under variable conditions. As an experiment, the GLASS albedo algorithm has successfully achieved an integration of multiple algorithms. It is anticipated that in the future, more and more product generations will rely on assembling multiple algorithms.

As more land products are produced and validated, our knowledge of the spatial and temporal variations of many variables is expanding rapidly. Such a priori knowledge should be objectively incorporated into satellite inversion algorithms to improve the inversion accuracy. Inversion of land surface parameters is generally a nonlinear ill-posed problem, and use of regularization methods incorporating a priori knowledge and integrating multiple-source data from different spectra and instruments deserves further research.

S. Liang et al., *Global LAnd Surface Satellite (GLASS) Products*,
SpringerBriefs in Earth Sciences, DOI: 10.1007/978-3-319-02588-9_7,
© The Author(s) 2014

Given the proliferation of multiple land products, quantifying the accuracy of each is critical. Creating more ground measurement networks under different conditions through international coordination and collaboration will help ensure the high-quality validation of various products. Scaling "point" measurements to satellite pixel scales using multiple sources of remotely sensed data at different spatial resolutions is also an important and essential research direction to pursue.

Appendix A
GLASS Product Archive and Distribution

The GLASS products are available at

- the Beijing Normal University Center for Global Change data Processing and Analysis at http://www.bnu-datacenter.com
- the University of Maryland Global Land Cover Facility at http://glcf.umd.edu.

S. Liang et al., *Global LAnd Surface Satellite (GLASS) Products*,
SpringerBriefs in Earth Sciences, DOI: 10.1007/978-3-319-02588-9,
© The Author(s) 2014

Appendix B
Acronmys

AB	Angular Bin
ADEOS	The Advanced Earth Observing Satellite
AERONET	AErosol RObotic NETwork
ANN	Artificial Neural Networks
AOD	Aerosol Optical Depth
ASTER	Advanced Spaceborne Thermal Emission and Reflectance Radiometer
AVHRR	Advanced Very High Resolution Radiometer
BBE	Broadband Emissivity
BELMANIP	Benchmark Land Multisite Analysis and Intercomparison of Products
BF	Baseline Fit
BRDF	Bidirectional Reflectance Distribution Function
BRF	Bidirectional Reflectance Factor
BSA	Black-sky-albedo
CALIPSO	Cloud-Aerosol Lidar and Infrared Pathfinder Satellite Observations
CASA	Carnegie-Ames-Stanford Approach
CC	Correlation Coefficient
CCCM	CALIPSO, CERES, Clousat, and MODIS
CDR	Climate Data Record
CEOS	Committee of Earth Observation Satellite
CERES	Clouds and Earth's Radiant Energy System
CIMEL	Ciencia e Investigación Médica Estudiantil Latinoamericana
CIMSS	Cooperative Institute for Meteorological Satellite Studies
CLM2	Community Land Model Version 2
CMG	Climate Modeling Grid
CNES	Centre National d'Etudes Spatiales
DOY	Day Of Year
EC-LUE	Eddy Covariance-Light Use Efficiency
ESDR	Earth Science Data Record
ESU	Elementary Sampling Unit

S. Liang et al., *Global LAnd Surface Satellite (GLASS) Products*,
SpringerBriefs in Earth Sciences, DOI: 10.1007/978-3-319-02588-9,
© The Author(s) 2014

EUMESAT	European Organisation for the Exploitation of Meteorological Satellites
FPAR	Fraction of absorbed PAR
FSW	Fixed Swath Width
FTIR	Fourier Transform Infrared
GCTP	General Cartographic Transformation Package
GEWEX	Global Energy and Water Exchanges Project
GLASS	Global LAnd Surface Satellite
GLO-PEM	Global Production Efficiency Model
GMAO	Global Modeling and Assimilation Office
GMS	Geostationary Meteorology Satellite
GMTED	Global Multi-resolution Terrain Elevation Data
GOES	Geostationary Operational Environmental Satellite
GPP	Gross Primary Production
GRNNs	General Regression Neural Networks
HPC	High-Performance Computing
HRVIR	Visible and Infrared High-Resolution
IGBP	International Geosphere–Biosphere Program
ISCCP	International Satellite Cloud Climatology Project
ISIN	Integerized Sinusoidal
ISR	Incident Shortwave Radiation
ISSTES	Iterative Spectrally Smooth Temperature and Emissivity Separation
LAI	Leaf Area Index
LSE	Land Surface Emissivity
LST	Land Surface Temperature
LTDR	Long-term Land Data Record
LUT	Look-up Tables
MDBA	Murray-Darling Basin Authority
MERIS	Medium-Resolution Imaging Spectrometer
MFG	Meteosat First Generation
MISR	Multi-angle Imaging SpectroRadiometer
MODIS	Moderate Resolution Imaging Spectroradiometer
MODTRAN	Moderate Resolution Transmission
MOR	Method of Ration
MSG	Meteosat Second Generation
MTSAT	Multifunctional Transport Satellites
MVC	Maximum Value Composite approach
NAALSED	North American ASTER Land Surface Emissivity Database
NAO	North Atlantic Oscillation
NASA	National Aeronautics and Space Administration
NCAR	National Center for Atmospheric Research
NDHD	Normalized Difference Between Hotspot and Darkspot
NDVI	Normalized Difference Vegetation Index
NEE	Net Ecosystem Exchange
NH	Northern Hemisphere

NOAA	National Oceanic and Atmospheric Administration
NPP	Net Primary Production
NSA	Northern Study Area
PAR	Photosynthetically Active Radiation
PARASOL	Polarization and Anisotropy of Reflectances for Atmospheric Sciences coupled with Observations from a Lidar
PNNs	Probabilistic Neural Networks
POLDER	Polarization and Directionality of the Earth's Reflectances
QC	Quality Control
RBFNs	Radial basis Function Networks
RMS	Root Mean Square
RMSD	Root Mean Square Deviation
RPV	Rahman-Pinty-Verstraete
RTMs	Radiative transfer models
SCE-UA	Shuffled Complex Evolution Method developed at the University of Arizona
SDSs	Separate Scientific Data Sets
SEVIRI	Spinning Enhanced Visible and Infrared Imager
SH	Southern Hemisphere
SPOT	Satellites Pour l'Observation de la Terre or Earth-observing Satellites
SRB	Surface Radiation Budget
STF	Statistics-based Temporal Filtering
SURFRAD	Surface Radiation
TES	Temperature and Emissivity Separation
TIGR	Thermodynamic Initial Guess Retrieval
TIR	Thermal Infrared
TOA	Top-of-atmosphere
TOMS	Total Ozone Mapping Spectrometer
UTM	Unified Threat Management
UWIREMIS	University of Wisconsin Global Infrared Land Surface Emissivity Database
VALERI	Validation of Land European Remote sensing Instruments
VNIR	Visible and Near-infrared
WMO	World Meteorological Organisation
WSA	White-sky-albedo
XRD	X-ray Diffraction